Praise for *The Lean Farm*

"Farming is not *just* a business, but it's still a business, and Hartman's application of Toyota's efficiency principles to the farm is nothing short of profound. As I read this fantastic book, my mind literally skipped from procedure to place on our farm with new ideas on how to create efficiencies. *The Lean Farm* should be dissected, digested, and discussed—then applied—on every single farm: big or small, wholesale or retail, livestock or produce. It would make all farms more profitable, productive, and pleasurable."

—JOEL SALATIN, owner of Polyface Farm,
author of *You Can Farm* and *Folks, This Ain't Normal*

"Anyone who thinks lean is only for a factory should read this book. Ben Hartman, with simple but eloquent prose and delightful figures and photos, demonstrates how all aspects of lean can apply to farming, a process of growing and selling living things. The mysterious uniqueness of farming under constantly changing conditions became clear as Ben learned to understand his customers and his value streams to increase value and eliminate waste. And lean reinforced, rather then replaced, the strong social values of the Hartman farm."

—JEFFREY LIKER, author of *The Toyota Way*

"Farmers are good at farming—it is what they enjoy doing! At the same time, planning, organizing, and working out everything most efficiently is often not done as easily. *The Lean Farm* will help us all easily increase flow, production, and income. It is a treasure trove of possibilities without the need for increased investment!"

—JOHN JEAVONS, author of *How to Grow More Vegetables*,
executive director of Ecology Action, and developer of
sustainable, biologically intensive mini-farming

"If you want to see, right now, what food farming will look like in the coming years, this is the book for you. Using the kind of super-efficiency that new-age manufacturing has perfected, author Ben Hartman describes, in great detail and with superb illustrations, how he and his wife reduced their farm size from three acres to one and still make a decent living on it."

—GENE LOGSDON, aut   *Farmer*

"Ben Hartman is diversified farming's Dean of Lean. He walks the talk, sharing insights on how lean principles helped his farm and how they can help yours. 'Lean' is the epitome of efficiency, an essential ingredient of any successful farm."

—RICHARD WISWALL, author of
The Organic Farmer's Business Handbook

"With lean principles, what's good for the farm is even better for the farmer. As we invite new farmers back to the land, into vacant lots, and onto rooftops, we have to give them the tools for success and the ability to sustain. 'Lean farming' won't leave you trying to turn a farm into an automotive factory, but you will get a whiff of what it means when the rubber hits the road."

—PHILIP ACKERMAN-LEIST,
author of *Rebuilding the Foodshed*

"We give every new employee a copy of Ben's writing to study. Adopting lean principles has been critical for bringing organization, focus, and harmony to our 100-acre fully diversified vegetable farm. 'A place for everything, and everything in its place' is a refrain we repeat over and over."

—PETE JOHNSON, organic farmer and
owner of Pete's Greens, Craftsbury, Vermont

"Clay Bottom Farm is a gem of a place in northern Indiana, where we are repeatedly told that you need a thousand acres to make a living as a farmer. Ben Hartman and his wife Rachel disprove this 'conventional wisdom' every day by managing a thriving farm business, not on a thousand acres, but on just one. In *The Lean Farm*, Ben explains how their elegant approach can be applied by anyone. His writing, like his farm, is clean, well organized, and easy to follow—but his ideas are revolutionary. *The Lean Farm* is one of the most original and innovative books on food and farming to come out in the last decade."

—STEVE HALLETT, professor of horticulture, Purdue University,
and author of *Life without Oil* and *The Efficiency Trap*

The
Lean
Farm

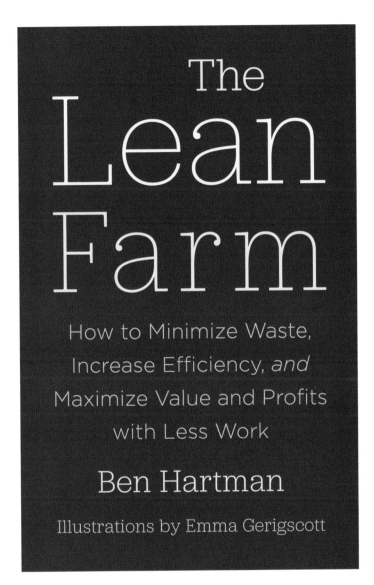

The Lean Farm

How to Minimize Waste,
Increase Efficiency, *and*
Maximize Value and Profits
with Less Work

Ben Hartman

Illustrations by Emma Gerigscott

Chelsea Green Publishing
White River Junction, Vermont

Project Manager: Bill Bokermann
Project Editor: Benjamin Watson
Copy Editor: Alice Colwell
Proofreader: Helen Walden
Indexer: Lee Lawton
Designer: Melissa Jacobson

Printed in the United States of America.
First printing August, 2015.
10 9 8 7 6 5 4 3   16 17 18

Our Commitment to Green Publishing
Chelsea Green sees publishing as a tool for cultural change and ecological stewardship. We strive to align our book manufacturing practices with our editorial mission and to reduce the impact of our business enterprise in the environment. We print our books and catalogs on chlorine-free recycled paper, using vegetable-based inks whenever possible. This book may cost slightly more because it was printed on paper that contains recycled fiber, and we hope you'll agree that it's worth it. Chelsea Green is a member of the Green Press Initiative (www.greenpressinitiative.org), a nonprofit coalition of publishers, manufacturers, and authors working to protect the world's endangered forests and conserve natural resources. *The Lean Farm* was printed on paper supplied by QuadGraphics that contains at least 10% postconsumer recycled fiber.

Library of Congress Cataloging-in-Publication Data
Hartman, Ben, 1978- author.
  The lean farm : how to minimize waste, increase efficiency, and maximize value and profits with less work / Ben Hartman ; illustrations by Emma Gerigscott.
    pages cm
  Other title: How to minimize waste, increase efficiency, and maximize value and profits with less work
  Includes bibliographical references and index.
  ISBN 978-1-60358-592-7 (pbk.) — ISBN 978-1-60358-593-4 (ebook)
1. Agriculture—Waste minimization. 2. Lean manufacturing. I. Title. II. Title: How to minimize waste, increase efficiency, and maximize value and profits with less work.

 TD930.H37 2015
 628'.746—dc23
                    2015015957

Chelsea Green Publishing
85 North Main Street, Suite 120
White River Junction, VT 05001
(802) 295-6300
www.chelseagreen.com

Contents

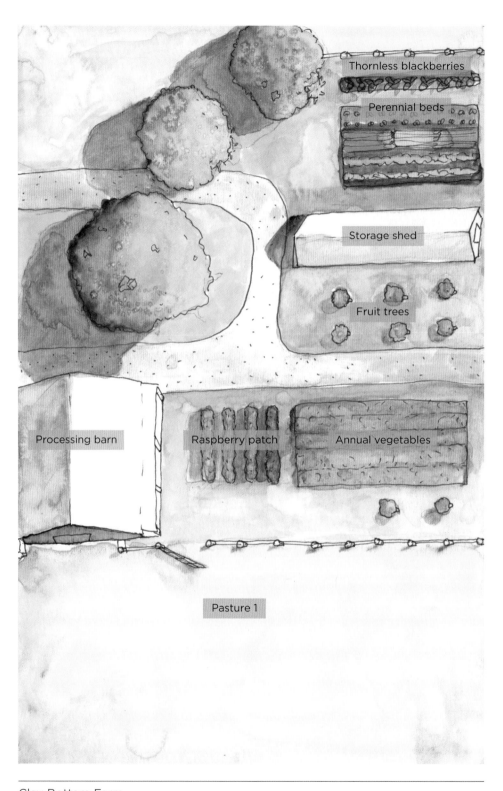

Thornless blackberries

Perennial beds

Storage shed

Fruit trees

Processing barn

Raspberry patch

Annual vegetables

Pasture 1

Clay Bottom Farm.

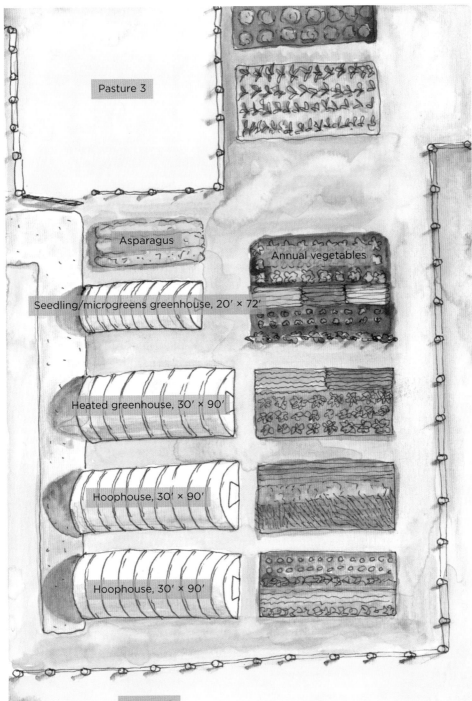

Pasture 3

Asparagus

Annual vegetables

Seedling/microgreens greenhouse, 20′ × 72′

Heated greenhouse, 30′ × 90′

Hoophouse, 30′ × 90′

Hoophouse, 30′ × 90′

Pasture 2

Acknowledgments

This book would not have been possible without contributions from many people. First, I thank Steve Brenneman, CEO and founder of Aluminum Trailer Company, Nappanee, Indiana, and a Clay Bottom Farm customer, who introduced us to lean concepts and who generously consulted with us for several years as we steadily implemented lean on our farm.

In October 2014 I met in Chicago with Susanne Pejstrup, a lean coach from Denmark who for many years has worked with farmers in Scandinavia, other parts of Europe, and Asia to implement lean systems on their farms. She described to me lean systems in use on these farms and contributed photos that enrich this book. Thank you.

Likewise, I thank the farmers who contributed photos and allowed me to interview them about how they employ lean: Pete Johnson at Pete's Greens, Craftsbury, Vermont; Randy Ewert at Bair Lane Farm, Marcellus, Michigan; and Steve Lecklider at Lehman's Orchard, Niles, Michigan, among others. The diversity and success of their operations is living proof that lean concepts can benefit *all* types and sizes of farms.

I thank Lynn Byczynski, author and editor/publisher of *Growing for Market* magazine, for her encouragement to publish this book. Her indispensable magazine is a constant drumbeat of improvement ideas, many of which we have implemented on our farm.

Some people claim that the lean production system is the most powerful production system in the world. Fortunately, several scholars have done the careful work of distilling the most basic concepts and putting them into words that everyday managers can use. In particular, this book relies heavily on the work of James P. Womack, president and founder of the Lean Enterprise Institute (Brookline, Massachusetts); Daniel T. Jones, chairman and founder of the Lean Enterprise Academy (Goodrich, UK); and Daniel Roos, Japan Steel Industry Professor of Engineering and founding director of MIT's Engineering Systems Division. Perhaps the most inspirational (and literary) writing on the subject I have found are original works by Taiichi Ohno, the Chinese-born founder of the Toyota Production System, who died in 1990 in Japan. Ohno's short but wise books, now translated into

English, contain no-nonsense headings like "If You Are Wrong, Admit It" and "Look Straight at the Reality" and continue to inspire firms—and farms—all over the world to root out waste and think more creatively.

I must also thank the dozens of interns and workers who over the years have contributed to the continuous improvement of our own microfarm. Many, many ideas for doing things more efficiently at Clay Bottom Farm are not my own but theirs. Thanks especially to Emma Gerigscott for contributing illustrations.

I am grateful, too, to Ben Watson and the team at Chelsea Green Publishing, who have been professional, supportive, and helpful. Ervin Beck, professor emeritus of English at Goshen College, offered shrewd editorial advice.

My wife, Rachel, also read this manuscript many times and leaned it up. Editing is a thankless task, and this book is as much the work of her insight as it is my writing. It wouldn't be in your hands if it weren't for her work and support. To my life and work partner, thank you.

Introduction

A few years after my wife, Rachel, and I took the leap and started farming full time, one of our customers, Steve Brenneman, a lean coach and the CEO and founder of an aluminum trailer manufacturing company, offered to come to our farm, watch us work, and talk with us about ways to "lean up" our operation.

The basic goal of lean production is to ruthlessly eliminate waste—anything the customer does not value—from your production system; essentially, you seek to create your product with as few interruptions as possible in the flow of work. The concepts behind lean originated in Japan and are now widely used in all types of industries around the world. Toyota in particular is a model of lean.

At first I was skeptical. Lots of well-meaning folks regularly give us farming advice, and we can't incorporate every idea that comes our way. Did lean principles really have a place on a farm? Aren't factories and farms very different places? Did I really want to turn our organic vegetable farm into an assembly line? Besides, what does manufacturing a trailer have to do with growing a tomato?

As it turns out, plenty.

◆ ◆ ◆

Our farm at the time was in a state of flux. As with all new farms, the first few years had been a flurry of high-energy experimentation and construction. We had built greenhouses, a walk-in cooler, and a processing room; dug trenches for drain tiles and water lines; bought a tractor and implements, a skid loader, and a John Deere Gator, along with a slew of hand tools, some of which we found useful and others that we discarded in various corners of the farm.

In just a few growing seasons, we had tested out hundreds of crop varieties and had experimented with growing techniques ranging from cutting-edge greenhouse practices to age-old methods gleaned from our Amish neighbors.

We were making it, but workdays were long, leisure time short. Some days we worked from sunup to well past sundown and still had supper to prepare. We hadn't been on a vacation in several years, partly because we were reinvesting all profits in the farm and partly because we felt that we couldn't leave. What if the greenhouses overheated? Who would take care of a sick animal while we were gone? We didn't have training systems or standards in place that would allow other people to perform even simple tasks that were needed to keep the farm going in our absence. Our production, on the whole, was erratic: every week we seemed to seesaw between overproducing and underproducing. We had a sense that if our farm was to survive for the long haul, the chaos would need to settle down.

So we decided to give lean a try.

▦　▦　▦

We started with a tour of Brenneman's factory. He showed us that instead of a centralized tool storage area, their workers used human-scaled, customized tool carts on wheels. *No matter the task, the right tool was always within arm's reach.*

The tool carts utilized a color-coded two-bin system for parts replacement: when one bin of bolts was empty, the worker placed the empty,

labeled bin on top of the cart and kept working from the second, identical bin. Twice a day one person walked around the factory, collected the empty bins, and replaced them with full ones from a room where parts were meticulously labeled and tracked. *Supplies were always ready in the right place, at the right time, in the right amount.*

Pea shoots sprouting on our farm. Photo by Emma Gerigscott/Clay Bottom Farm.

The workspace was well lit and remarkably clean. It felt *good* to be

Lean offers a management system to help farmers analyze their work. Photo by Emma Gerigscott/Clay Bottom Farm.

there. These were not the dim, dirty environs one pictures on hearing the word "factory." *Organizing and cleaning were part of workplace culture.*

The workers were skilled and well trained. On the wall next to workstations were clear outlines of worker tasks. The outlines explained each procedure and gave reasons for the task. Brenneman told me that with their training protocols, "even complex tasks can be taught to nearly anyone." *Jobs were broken down into steps that could be easily learned, and standards were clearly visible.*

After the factory tour Brenneman showed me the office spaces for the accountants, engineers, and executives. I was surprised: for a manufacturing business of this size I expected corner windows, flashy paperweights, and Herman Miller chairs. Instead, the corporate offices, carved into a corner of the factory building, were modestly sized with adequate but certainly not extravagant furnishings. *Even office spaces had a feeling of efficiency and focused production.*

Most important, Brenneman designed his factory with his customers in mind. "A few years ago we were discussing the possibility of new offices. It would have been an expensive project. Then we asked ourselves, from a lean standpoint, *What do our customers want?* We decided our customers didn't really care what our offices looked like. What they wanted were quality trailers. So that's where we put our resources."

We were hooked. What if we could apply the lean principles that worked so well in a factory to our organic farm? Not all principles would translate perfectly, but what would happen if we tried out a few?

Farms and factories are very different places, but in the end our task is the same: to deliver a high-quality product to customers who value what we make or produce. We began imagining our tools neatly lined up in tool stations, close to their point of use. What if there was no clutter, no stumbling over crates? What if every item on the farm was well cared for? What would happen if we really took seriously what our customers valued? How would that change how we washed lettuce or designed our processing area? What if no movement was wasted, and every single seed turned into a product that sold?

We started to see how lean might give a boost to our small farm without upending our values. We had something to strive for, a new vision, and new energy to meet the challenge.

A Basic Definition of Lean

In its simplest form, Taiichi Ohno, the founder of the Toyota production system, described Toyota's lean method as "looking at the time line from the moment the customer gives us an order to the point when we collect the cash. And we are reducing that time line by removing the non-value-added wastes."[1]

Another way to view lean is by analyzing capacity, the amount of product that can be produced in a given span of time. To Ohno, the capacity equation is simple:

"Present capacity = work + waste."

The lean way to increase the capacity is to eliminate the waste. "Work" is anything that adds value for the customer; "waste" is anything that doesn't. The eventual goal is zero waste and 100 percent work.[2]

This "absolute elimination of waste" became the backbone of the Toyota production system, and it catapulted the company in the latter part of the 20th century past its rivals to become the largest automobile manufacturer in the world. It was also the most profitable: by the early 2000s Toyota's net profit margin was 8.3 times higher than the industry average.[3] The very simple formula was to find waste, root it out, and turn it into capacity to produce more.

Heirloom tomatoes. The goals of lean are to cut out waste and to find and deliver value.

In their book *Lean Thinking*, James P. Womack and Daniel T. Jones define the lean approach as a set of steps, arguing that lean boils down to five principles:[4]

1. Precisely specify what customers *value.*
2. Identify the *value stream* for each product.
3. Make value *flow* without interruptions.
4. Let the customer *pull* value from the producer.
5. Pursue *perfection.*

Here is how these principles apply to a farm:

1. *Precisely specify value:* Dig deep and really listen to your customers. Scratch below the surface. Research what really gets your buyers excited. The precise good or service your customer wants should drive your farming.
2. *Find the value stream:* Once you know what customers value, map out your farm and trace that value—your product—as it's being created.

French breakfast radishes.

Don't skip steps. Start your map at the beginning, with planning and ordering seeds, and finish with depositing cash in the bank.

3. *Create flow:* Now scour your map—your farm—for waste, anything that doesn't add value. Once you find it, plan to root it out. The goal is to make value *flow* without interruption.

4. *Sell through pull:* Instead of growing too many radishes or raising too many chickens and *pushing* the excess, produce exactly what customers want, in the amounts they want, when they want it. Let their *pull* guide your production.

5. *Aim for perfection:* Develop a farm culture of continuous improvement with the goal of achieving perfect flow: zero waste production.

This book is the story of how we used these ideas on our microfarm to grow and sell more food than we ever thought possible.

Factory versus Farm

In some circles, to pair the terms "factory" and "farm" is sacrilege. Indeed, many of the dire problems facing our environment in the 21st century come from the inappropriate application to the family farm of factory-style production methods and output standards. When yield per acre or gallons per cow is your only standard of success, when living animals are "production units" and produce or grain fields are nothing more than "profit centers," then the mystery of life is overlooked and the health of the planet suffers.

There are, of course, important differences between factories and farms, and these affect how lean should be used on farms. The raw materials of industry are inert and predictable. Steel, aluminum, wood, and copper are molded, shaped, fabricated, and assembled to form objects for sale. In farming our raw materials—our seeds, plants, and animals—are constantly changing form. Even soil is constantly changing. Researchers at Oregon State University determined that a single teaspoon of garden soil contains up to a billion bacteria, along with "several yards of fungal filaments, several thousand protozoa, and scores of nematodes."[5] These are all *living* soil organisms. Farmers cannot predict with precision how these organisms will interact with tomato plants or blueberry bushes or potatoes.

Imagine running a factory without a roof and you have a picture of the vulnerable and dynamic nature of a farmer's work. Weather shifts, which can occur several times a day, force farmers to adapt plans repeatedly. Heavy rains, strong winds, and temperature swings will leave a manufacturing facility largely unaffected but can destroy an unprotected crop of lettuce, costing the farmer several months' worth of wages. We head to town on a cloudy morning, confident in the prediction of overcast skies, and watch the clouds roll away. Do we then make the twenty-minute drive back home, abandoning our errands, or just stay where we are and hope for the best in the quickly overheating greenhouses?

In his book *Oil and Honey*, author and activist Bill McKibben writes of learning through a beekeeping friend how keeping bees is like playing a game of chess, where with each move—each day of unpredictable weather, each small fluctuation in how things are growing—the entire board is different.[6] Good farmers are keen observers of nature, adjusting their moves and their strategies every day.

As Wendell Berry says, "The farmer lives and works in the meeting place of nature and the human economy."[7] Farm products will always be intimately tied to ecosystem behavior in a way that factory goods are not.

 ▪ ▪ ▪

Climate change is an undeniable threat to our farm and our livelihood. In the winter of 2013–2014, we managed to survive multiple polar vortexes, and in fact we continued harvesting carrots, spinach, kale, and other cold-hardy crops through many nights below 15 degrees F (-9 degrees C). I still remember running up to the processing room from our greenhouses with totes of spinach and salad greens in order to keep the crops from freezing en route. We relished every leaf in those extremes, but our yields were low and our income suffered. Is our food system more vulnerable because of climate change? No doubt.

With weather patterns becoming more and more erratic, agricultural climate control becomes increasingly important. Our greenhouses got us through that arctic winter. Irrigation systems in the Midwest, used to combat drought, have spread in the past few years like an octopus's tentacles across field after field. Animal buildings and greenhouses rely on more and more robust ventilation and heating systems—anything that can provide a buffer between our products and raw nature.

While new technologies for climate control suggest that the worst climate swings can be mitigated, completely shutting farming off from nature—if that were even possible—would not be true farming. Industrial farmers, perched in the cabs of air-conditioned tractors, distance themselves and their work from nature, to the peril of us all. Carbon-heavy, production-at-all-costs agriculture produces greenhouse gases at an alarming rate; it turns fertile ground into wastelands; and it pollutes the water and air we all rely on. It is ultimately self-defeating: it gobbles good topsoil without replenishing the land; it robs from the fossil fuel bank without giving back; and it contributes to erratic weather that will eventually make farming an impossible task. It is a style of farming that cannot endure because it misinterprets the work of farming.

A farmer's work is more like that of a horse trainer than a mechanic, more like that of a healer than a computer repairperson. It is not really accurate to say that farmers grow food or raise animals. Farmers alter environmental conditions in such a way as to maximize a plant's or an animal's innate ability to do its own growing—in the same way that the best horse trainers seek to draw out abilities already *within* their horses or in the way the best healers know when to stand back and let their patients' bodies do the work. There is mystery in farming. While there is beauty in the craft of shaping inert materials into useful things, it is a different type of work from tending to the living beings of a farm.

Some 12,000 years ago humans entered into a new type of relationship with nature. No longer were we content to glean berries and hunt game. We learned ways to participate even more fully in the cycle of life by altering nature's "raw materials," its flora and fauna, to feed ourselves. While lean methods can make that work more efficient, they should not be used to completely remove nature from farming, even if less nature sometimes means easier, more profitable farming. A lively and dynamic nature is both the core challenge and the core ingredient of farming.

When farmers apply lean with the right intentions—to restore the earth and increase the health of families and local communities—their farms can produce lots of food *and* fall into alignment with nature. The number of production challenges facing manufacturers is surely equal to the number facing farmers. They are just different kinds of challenges. Appreciating the differences will help farmers use lean to its fullest.

Lean down on the Farm

After touring Brenneman's factory, we continued reading about lean principles and soon began applying them on our farm.

We currently farm a 5-acre property, with less than 1 acre in production, including 9,000 square feet (roughly ⅕ acre) under four greenhouses. My wife and I farm year-round and earn a comfortable income from farming. One part-time staff person helps us throughout the year, and from March to November one or two additional interns join our team. Our microfarm is

A greenhouse ready for plants. Lean says to clear out clutter so you can see your work.

one of thousands such farms that now dot the landscape. Large farms may dominate US agriculture, but small-scale farming seems to be "happening everywhere, including in cities and suburbs," as Lauren Markham, a writer for public radio's *This American Life*, explains. This new "ubiquity of agriculture" gives consumers access to goods produced on a small and local scale in ways that they didn't have before.[8] We are a tiny operation compared to many who use lean principles, but we've found that lean works well on our scale.

We sell roughly equal amounts to three outlets: to five area restaurants and a grocery store; to a farmers' market, where we participate in a collective community-supported agriculture (CSA) program; and to around sixty of our own CSA customers, who pay us at the beginning of the season and receive our food nine months out of the year.

We did not arrive here overnight. For five years, while we were in our twenties, we lived in town, farmed part time on rented land, and worked part time at other jobs as we saved for a down payment on our current farm, where Ben has farmed full time for another six years. For four of those years, Rachel continued working in the public schools. We grew gradually and steadily: our first CSA had only twelve members, then twenty, then thirty-five, and now sixty. We've kept our equipment simple and to a small scale and have focused on increasing our knowledge and improving our process.

Our savings helped us take off quickly when we purchased our current property. A large down payment meant that our monthly costs were

Lean to Keep Farms Small—and Viable

Lean is so effective as a tool that it might be tempting to use it as a means to get big. But lean can be used equally well to increase profits while keeping your farm small.

The dominant business model in the United States assumes a successful business will grow in size every year. But can one also find one's right size and be satisfied with it? Lean principles—which focus on creating more value over simply more quantity—opened up the possibility for us of *not* growing bigger, of thriving even within the constraint of size.

In fact, scaling *back* might sometimes be the right decision for a farm. With lean as a guide, we've steadily shrunk our growing area to less than an acre, and we've increased our profits every year. If you can cut back in the size of your production area and—by eliminating waste and working out process kinks—still net a comfortable salary, why not do so? Through the steady application of lean principles—systematic waste elimination, a focus on crops with high margins, careful planning to harvest crops when they are in highest demand—our farm has stayed small and profitable. Lean for us means doing more and more with less and less.

In the United States, profit-through-growth farming is more common. Huge, industrial-scale agriculture—which serves the interests of investors and agribusinesses, not local communities—dominates the nation's rural landscape. According to a USDA report, farm size has doubled since 1990 "and the trend is likely to continue."[9] A recent article from the *Washington Post* states bluntly: "Farms are gigantic now. Even the 'family-owned' ones."[10]

The problems with supersizing farms are many. Rural communities are left tattered. Many farms are sterile, uninviting places void of windbreaks, trees, and habitats for snakes, dragonflies, and migratory birds. In 1900 a majority of farms raised chickens, cattle, milk cows, and hogs in addition to a wide range of fruits and vegetables. Now just a tiny percentage can claim that diversity. Farms used to be places that connected us to nature, where children played and learned about the world. Now many megasized farms so little resemble the natural world that children would be lost on them.

Fortunately, it is also true that small-size farms are also on the rise. Lean can help the small producer survive, as I hope this book shows, by cutting costs and harnessing the *advantages* of small. Small-scale farmers who apply lean principles can serve their customers better than do gigantic producers and at the same time produce in sufficient volume to support their farms.

reasonable and that we could focus our efforts on farming. Except for a small family loan early on to help put up a greenhouse (which we quickly paid off), we've managed to pay for capital improvements as we've gone along rather than racking up debts and interest payments.

We've had good teachers along the way. We both worked for Kate and James Lind at Sustainable Greens in Three Rivers, Michigan, where we honed our skills. The Linds, along with their son, Matt, were pioneers in supplying fresh greens to fine-dining Chicago restaurants long before the winter greens movement took off. We also owe plenty to creative direct-market pioneers like farmer and author Eliot Coleman and to our Amish and other Plain neighbors, who for decades have been perfecting—and leaning—their craft.

LEAN AS INTUITION VERSUS LEAN AS SYSTEM

Lean habits come naturally to many farmers, especially those who sell directly to customers and who are intimately involved in their farm's production.

One idea that came naturally to us, for example, was the principle of inventory reduction. Lean manufacturers keep inventories low—they assemble cars on order rather than on speculation, for instance—because it costs money to warehouse goods that could become obsolete. On our farm, we've never had the luxury of filling warehouses or parking lots with large inventories of goods for months on end because much of our food will hold its value for about a week before it turns into mushy waste. While many manufacturing firms struggle to lean up by culling bad habits of inventory buildup, we were able to apply this principle quickly to improve our farm.

No farmer likes to see waste. Steve Lecklider, owner of Lehman's Orchard in Niles, Michigan, intuitively implemented many lean techniques on his farm because they made sense for his operation. He organized his 55-acre farm to operate in ways that maximized his efforts. He told me, "You can say this or that is a Japanese theory, but for farmers like us, we live this stuff every day. We're the ones who have to deal with the overproduction and all the waste. We learn this stuff by doing it." For farmers like Lecklider, waste is visceral—a pile of apples not nice enough to sell or milk that has no market.

Still, beyond intuition, what lean offers is a complete *system* for more efficient production tested over many years in many industries. The advantage of a system, especially one as thoroughly vetted as lean, is that its effects can be more wide-reaching than the application of any single principle on its own. The lean system gives farmers a place to start as well as a set of best practices to refer to.

LEAN BUT NOT MEAN

The lean system is often misunderstood. Lean farming does not mean bare-bones production and meager harvests. It does not mean stingy farmers and barely fed cows. It does not mean dull and uninteresting work or working faster and harder to get jobs done in less time.

Rather, lean makes work easier to do and more meaningful. It pivots your farm toward value—trimming away activities and things that do not contribute to value creation so that all efforts count. Lean farms are productive and abundant, vibrant and creative.

One of the reasons we hesitated to start applying lean was that we liked what we did. Farming was fun, even if it was at times chaotic. We didn't want a system to take the joy out of our work. That hasn't happened. Instead, lean principles have helped us clean up our farm so it is now a more pleasant place to work, and we are able to do *better* work. Lean has given us tools to guide our everyday production and our long-term decisions. We've learned how to better engage our workers so we are all working toward collective goals. Lean has made us more profitable, giving us time to rest and conserve energy for other pursuits, like raising our young family and planning a little more travel.

▪ ▪ ▪

Sadly, too many farmers, especially those just starting out, are mired in hard work and saddled with debt. The beloved pursuit of Thomas Jefferson is more endangered now than ever.

It is not exclusively the fault of these farmers: high land prices and chronically low food prices have put this career out of reach for many young people who would otherwise make great farmers. Farming requires start-up money and skills that take years to learn. Many people beginning their work lives are simply not willing to risk devoting such a large chapter of their lives to honing a trade that so few have found profitable.

While the hurdles are real, farms of any size, age, and type that adopt a lean approach can find success. Real opportunity for new and small farms still exists. Success is possible, as I hope this book shows, through customer-focused production that incorporates the best ideas in business management with new, right-sized technology and efficient, intensive production methods. Because profit margins will still be thin, a solid understanding of costs and a commitment to trimming waste is a must. For profit margins that pay the bills and afford a comfortable living, growers have no choice but to adopt a lean approach.

Business fads resemble diet fads, but lean enjoys remarkable staying power: it resonates in a culture of excess and greed. Lean says that cutting waste is a legitimate way to grow a business. Farmers have been told for decades to get big or get out. Lean offers a better way.

Keeping Your Values: Lean as Just a Tool

You can apply the lean system without stripping values or the human element out of farming. Lean is about a pivot toward value and people—what you and your community value most should take center stage on your farm. Lean helps to identify and deliver that value, not replace it.

If profit is your only motive, then lean principles taken to an extreme could easily result in a fossil-fuel-dependent monocrop farm where only the most profitable production methods and crops survive and all others are axed—or it could be used to justify harsh treatment of animals, as in confining them in small areas in the name of efficiency. This type of farming might employ selective lean tools, but it does not represent the true spirit of lean, which is to farm *better*, not meaner.

There is plenty of room in lean to farm with values. Take mechanization. On our farm we enjoy the benefits of mechanization, though our equipment, like a compact tractor, is smaller than equipment found on many farms. There is a place in lean for farmers to choose even *less* mechanization as a value statement. The Amish neighbors that surround our farm forgo tractors altogether and choose lower-tech and even fossil-fuel-free farming as a value. Lean ideas add profitability to any type of farming operation, not just mass-production farms.

In fact, lean can be used to *strengthen* sustainability. After all, lean is about removing waste. You don't have to discard values like sustainability to embrace lean.

In Part I of this book, I explore broad lean principles, cite lean tools that we have implemented on our farm, and show how other farmers have put them to use as well. In Part II I offer specific lean tips for new farmers and discuss the limits of lean in the context of agriculture, as well as how lean can be used for more than profit. Anyone on any type of farm should find in these pages ideas worth implementing.

Factories and farms are several worlds apart. Farmers and plant managers don't often mingle. But when different experiences do meet, there can be fertile results. I hope this book gives you inspiration and tools to root out waste and find success on your own farm.

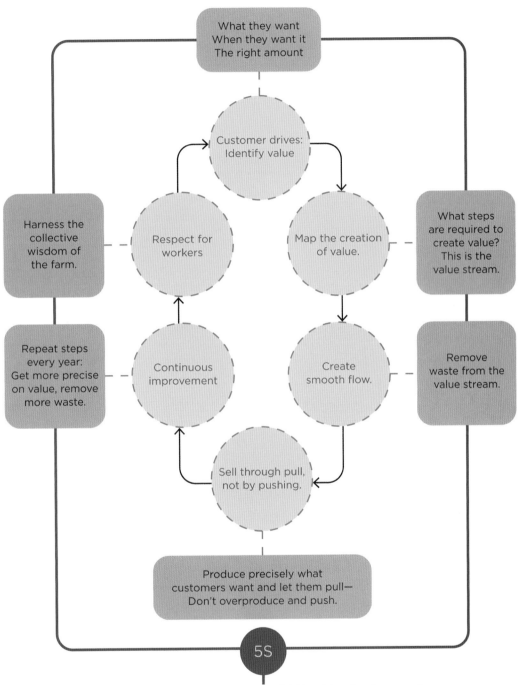

The Lean Farm Cycle. Lean ideas originated in Japan and are now used across a range of industries worldwide. They can help farms, too. This illustration is based in part on concepts from Taiichi Ohno, *Toyota Production System*, and James P. Womack and Daniel T. Jones, *Lean Thinking*.

PART I

Lean Thinking
on the Farm

CHAPTER 1

Every Tool in Its Place

So simplify the problem of life,
distinguish the necessary and the real.

—HENRY DAVID THOREAU

L ean's weapon for streamlining production environments is called 5S. Each practice in the system corresponds to a list of five Japanese words that start with the letter *s*: *seiri, seiton, seiso, seiketsu,* and *shitsuke*. In English: sort, set in order, shine, standardize, sustain.

Why start with 5S? Because it underlies the work of leaning a farm. Farms, like organisms, need to be flexible and adapt quickly to shifting environments. A bloated farm that is weighted down with junk and piles of out-of-date tools won't be able to move fast enough to keep up with the changing landscape of food production. A pretty farm is not the goal, though that might be a result. It has taken several years of repeated use of the 5S methods for us to see through the thick layers of waste that were slowing down our work.

The tasks of cleaning up and setting up organization systems might seem daunting, but let me explain 5S and break up the work into bite-size pieces.

1. *Seiri:* Sort

The first step is to ruthlessly eliminate anything that is not absolutely necessary for your production system. When in doubt, get rid of it. The only items left in your production spaces should be those you use every day to add value to your product. Period.

For us, this meant devoting several weeks of staff time to filling wagonloads for the auction and landfill. We hired a scrap metal hauler to take away rusted tools, mangled greenhouse parts, and thousands of bits of steel cluttering our workplace. In the first few years of our farm, we had tried out different seeders, hand tools, carts, and other types of equipment. It was time to liquidate what could not be justified. We breathed a sigh of relief each time a wagonload left the property. After those few weeks, a weight had been lifted: our farm was literally lighter, more nimble, and ready to produce.

SIMPLIFY

Much can be said for having exactly the right tool for a job. But this also can be a recipe for confusion and clutter. Often you are better off with fewer tools that accomplish a wider range of tasks. Rather than hoarding hundreds of specialized hoes, feeders, and implements, see what you can accomplish with as few tools as possible.

On our farm we got rid of our eclectic collection of harvesting knives and purchased one type of produce knife that worked well for almost all of our harvesting needs. The same with pruners: we hung onto identical curved grape shears and got rid of a basket containing twenty or more older models. As for equipment, we sold a quaint Ford 8N tractor and several older implements that duplicated tasks performed much faster with a newer Kubota tractor and PTO tiller.

We simplify every year, even as our business grows. A few months ago we sold a disc harrow and a rotary mower that we hadn't used in at least two years. We used the money to purchase a quick-hitch system that allows us easily to switch out the few implements that we *do* use. We sometimes change implements three or four times per day. The investment makes that task easier and saves several minutes each week, time we can use to grow more food. This is a concrete example of turning waste into capacity to do more.

The Cost of Keeping Everything

Everything you keep is a cost, even if the cost is not obvious. Every time you manage an item that is not part of your production system—every time you wash it, step over it, stash it away, even *think* about it—that item is taking an investment from you and your farm. It is costing time and money. Over time the costs add up. In fact, if you need proof, you can calculate the costs. Carry a stopwatch and turn it on every time you hunt for an item. Or calculate the costs of maintaining building space just so you can store unused items there. The results might surprise you.

OPEN UP THE VIEW

Getting rid of what you don't need has several advantages. First, staff have an easier time finding the tools you tell them to find. On many farms tools are so disorganized that the only person on the property who can retrieve the right item is the farmer, which wastes both the farmer's and the workers' time. This was our farm before we applied 5S.

Second, it is easier to take care of a few carefully chosen items rather than a barn full of random items. Is it faster to sharpen five hoes or fifty? To service one tractor or five? There is rarely a good reason to duplicate tools or machines or keep spares around.

Third, it is easier to see your product as it moves around your farm if there are fewer things clogging the production area. A farm should be so clean that observers can easily stand back and watch the flow of value as it is being created. They should see carrots being seeded, pulled, and washed or cows grazing on thick pastures—not piles of scrap metal or heaps of debris. Each screwdriver, bucket, hose, and tractor will take time to clean, stack, move, maintain, and manage. The fewer items you have, the easier it will be to keep the farm in order.

DESIGNATE A RED-TAG HOLDING AREA

Sometimes it is hard to identify unneeded items. In lean factories workers tag questionable items if they wonder:

Is the item really needed?
If so, is it needed in this quantity?
If it is needed, does it need to be located here?

After a week, enough time for managers to scan the room, the items are discarded, reduced in number, or moved to more appropriate locations on the production floor. We have set aside an area in our barn where we place marginal items. We call it our red-tag room. Rather than removing those items by the week, we wait for the annual spring or fall produce equipment auction held about an hour away. Any item left in the red-tag room is loaded up and taken to the sale.

Make no mistake, our farm still harbors many unused items. We still too often hang onto what we don't need or wait too long to sort. But by following a schedule and a process, we can make steady progress and get cleaner, not more cluttered, every year.

SORTING VERSUS LINING UP

Taiichi Ohno, the inventor of Toyota's production system, writes that many firms use concepts like *seiri* but few actually achieve *seiri*. He recounts his frustration with one Toyota factory:

> *I told them to sort and set in order, and when I visited them the next time, the parts were lined up in neat rows. I asked*

Hobbies, Prototyping, and Spare Parts

The more hobby and work areas are separated the better, mostly so that hobby items don't drift into the work area, and work items don't get lost in the fishing gear. This separation also makes finding items easier for workers.

Some items you might want to keep for experiments, not production. For those items, we built a wall of prototyping shelves. These store thermostats, heaters, electrical and plumbing parts, and other materials that we pull out for new projects. We make a distinction at sorting time between junk and an item of actual use for prototyping. Periodically, we scan the pro-

totyping shelves and quickly revisit our decision to keep each item in those spaces.

We also keep on hand a minimum number of spare parts, such as extra irrigation nozzles and backup heaters. Remember, anything you keep is a cost—in time to manage and in space to store the item. So we are careful to keep spares around only if the loss of the item for even a few hours could seriously harm production. We keep spares labeled and organized. We are happy to let the local hardware store keep extra shovels and wheelbarrows and other items that don't need to be replaced right away when they break.

*them, "Don't you know what it means to sort and set in
order?" Sort means to throw out what you do not need, as
when you do personnel adjustment. If you are holding on to
your parts and stacking them up in your warehouse just
because you worked hard to make them, this is not sorting.*[1]

Sorting is eliminating what you don't absolutely need, not displaying
your wares.

SORTING VERSUS SQUIRRELLING

It might be tempting, especially if your farm has extra buildings, to stow
away your unused items in spaces you feel are out of the way. Don't be
fooled. This is not sorting. This is squirrelling. The problem with squirrelling is that rarely does a farmer who uses the practice truly relegate all
unused items to "spare rooms." More likely, many of those items are
squirrelled away in recesses within the production space. Also, items
from the "squirrel room" often drift back into the work area and never
return home. And even if you can't see an item, it can still take up room in
your mind—you are still its owner, still responsible for it. There is no
exception to the rule that everything you keep is a cost, hidden though it
may be.

FINDING THE RIGHT DESTINATION
FOR UNUSED ITEMS

Sorting should be done with the same mindset of efficiency as other farmwork: usually, the faster you can dump waste from your property the
better. A common problem is hanging on to an item for too long because
you are waiting for the right moment or the right market to take action.
Maybe you're waiting for the perfect buyer to stumble onto your farm.
Don't. When you count time and hassle, it's best to get rid of unused items
as soon as you can. Holding out for a better price usually costs more than
the difference in price.

2. *Seiton:* Set in Order

Every tool on your farm should have a place. It should be in its place or in
the hands of a worker. There is no third option. Use the tool, then put it
back home.

STORE TOOLS WHERE YOU USE THEM MOST

When Brenneman first visited our farm, he asked if he could observe my tomato pruning process. So I walked about 100 yards to the storage room, found pruners, then trudged back to the greenhouse. I pruned a row of tomatoes, then headed back to the storage room to replace the pruners.

"Why not store your pruners in the tomato greenhouse?" he asked. The thought had never occurred to me.

Over the next year we proceeded to move almost all of our tools out of the tool shed and spread them around the farm, close to their point of use.

We now store hoses close to the plots that need irrigating. A high-powered magnet on the handle of our six-row seeder holds wrenches needed to adjust the seeder. We keep channel-lock pliers next to each hydrant in the greenhouses for attaching hoses. We store winter greenhouse row covers in the corners of the same greenhouses where they are used. One of our most-used tools is a rake for renovating beds of greens—we bought four of them and hang one by the entrance to each of our four greenhouses. When we need a tool, it is within easy reach rather than on the other side of the farm.

STORE TOOLS WHERE YOU SEE THEM

Our goal is to organize our tools in such a way that a ten-year-old can walk into a room and easily get us a shovel or a digging fork or whatever we've asked for. This meant investing in hooks or magnets so we could hang tools at eye level, within easy reach. Workers can now retrieve tools without our showing them where they are. And they know where to return them. Before, our practice was to go around the farm and clean up after everyone had gone home.

STORE ACCORDING TO FREQUENCY OF USE

Reserve your easiest-to-reach spaces for your most-used tools. For example, we use our harvest knives and pruners almost daily, so we store those tools prominently at eye level by the door to our biggest greenhouse and on the way in and out of our processing room.

Our leek dibbler, which we use once a year, is also at eye level but on a back wall, not in a prominent spot. The same applies to large equipment: the tractor and skid loader are housed by our processing area for easy access; the cultipacker and potato digger, which come out only a few times per year, sit farther away, at the edge of a field.

When I visited Brenneman's factory, he showed me their mobile tool carts, which are spread around the factory and replace a central tool area. The

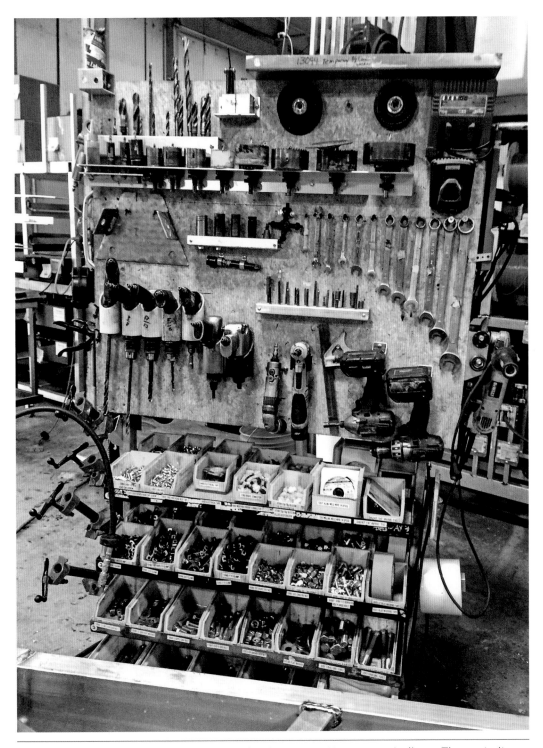

A mobile tool cart at the Aluminum Trailer Company, Nappanee, Indiana. The cart sits on casters, which allows workers to move it easily. The system keeps a wide array of tools always within arm's reach.

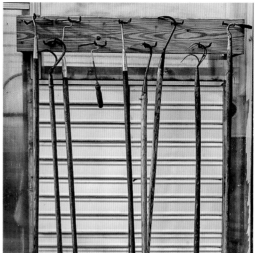

Before we applied 5S, all of our tools were housed in a central storage area, often in disarray. As part of our leaning, we got rid of unnecessary items and kept the small number of tools we used the most. We moved those tools out of the storage area and spread them around the farm close to their points of use.

carts are job-specific. For example, an electrician's cart contains every tool and part the electric installer needs. The carts all have casters, so there is no running from one end of a trailer to the other to retrieve a tool—you simply roll the cart along with you as you work. The carts are even equipped with outlets, and screw guns are permanently plugged in so that the right gun is always within arm's reach. Farmers would do well to place tools for efficiency of use, too.

Tools still get lost on our farm. We waste our share of time looking for the hoe that didn't make it back to its home hook. However, we have saved hours by thinking carefully about the right place to store the materials we need in order to keep our work flowing without interruption.

3. *Seiso:* Shine

Clean your workspaces with a toothbrush, then make sure they are well lit. Some lean factories aim to create workspaces that are as clean as a hospital. That might be unrealistic for a farm, but cleaning up spaces helps in several ways.

First, if a space starts out clean it motivates workers to keep it clean. One example is our propagation area. Before we implemented 5S, the area was always a mess. Dirty trays were stacked up haphazardly, cobwebs filled the corners, dirt accumulated on the floor and turned into mud. We spent

Before and after photos of our processing area. We applied 5S shine by painting the floor, hanging bright lights, and installing stainless steel equipment.

We sort and store tomatoes in an old milk-processing room. Lean says to illuminate workspaces so you can see waste.

an entire day one year scrubbing the space. The result is that while the space sometimes gets dirty, it has never descended back into chaos: we enjoy the cleaned space so much that we are motivated to keep it that way.

Second, it is easier to see what needs to be done if dirt and trash aren't in the way. In the field, for example, trash can obscure the view of our work. Old irrigation parts, used row covers, and empty seed bags lying on the ground clutter our landscape and confuse our workflow. The same thing happens in the processing room: piles of old boxes and dirty floors distract us from getting work done. Excess materials and junk are physical impediments—we waste time tripping over them—but just as importantly, they obfuscate our thinking: it's harder to plan our work when we don't have an uncluttered view of our workspace.

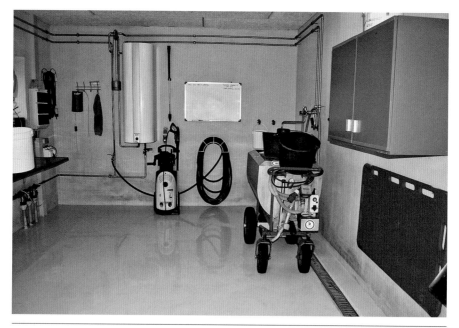

The 5S system was applied in this kitchen for feeding calves at Stensager-gaard dairy farm in Aars, Denmark. Photo by Susanne Pejstrup, Lean Farming®.

A clean farm is also safer. Injuries on many farms are the result of unclean, poorly lit spaces. Cleaned spaces make hazards easy to spot, as do bright lights—because we sometimes harvest at night, we installed flood-lights in all of the greenhouses.

And we shined in other ways: We painted the cement floor of our 12′ × 30′ processing area with epoxy paint so that we can mop up easily. Then we installed stainless steel counters and sinks that sanitize easily, and we hung four T8 fluorescent light fixtures. It might seem like overkill, but we think the increased quality of our work was worth the effort. A bright and clean work environment helps us see waste.

4. *Seiketsu:* Standardize

Sorting, setting in order, and shining need to be made routine. This is much easier when your systems are standardized and when the first three steps—sort, set in order, shine—are part of everyday work.

Standardizing 5S means that the same task should be performed the same way no matter who is doing the work. It means parking the skid loader in its home every day, replacing tools, and cleaning workspaces using the same sequence every time.

To standardize everyday tasks, we keep our systems as simple as possible. For example, we bought and placed around the property green-colored totes that are used exclusively for collecting weeds. The system is easy to follow, and having totes close to where the weeds are encourages us to keep our growing beds clean. We replaced our mix of thirty or so types of salvaged harvest containers of different colors and shapes with just two types of containers: fully enclosed and lidded Rubbermaid totes and vented bulb crates.

Instead of having growing areas of different sizes and shapes, we divided our farm into eight plots of the same length, and we use uniform growing beds in those plots. This means we can install uniform irrigation systems, plastic mulches, and drip tapes in the same way in any bed without confusion. For our market booth, we purchased display baskets of the same size and shape. These can be disinfected, and they nest inside each other when not in use.

SHORT AND OFTEN

Straighten-and-shine tasks are best accomplished in short steps done frequently, as part of day-to-day routines, rather than as time-gobbling chores set aside for once or twice a year. Too often farmers tell themselves they'll have time to clean up during the winter, while every day their farms become more and more inefficient because of clutter. This was certainly our habit.

A better practice is to put tools away after each task and to restore the farm to order each day. 5S tasks thus become short and manageable and don't pile up and become overwhelming. On dairy farms daily cleaning is not optional in milk-processing areas. But other types of farms would do well to implement daily cleaning regimens, too.

We do reserve certain maintenance tasks—oiling handles, sharpening blades, disinfecting greenhouse equipment—for the off-season, but in general we aim to keep our production areas consistently clean. It doesn't always happen, but we're getting better at it. Once again, it helps to have a target.

VISUALIZE IT

The management of sort, set in order, and shine is best done with visual commands rather than orally or only in writing. Visual system management (VSM) has any number of uses in lean (see Chapter 9). For 5S tasks, farms can "visualize" how to organize in many ways:

- Use shadow boards. Draw contoured outlines around tools to indicate a home for each.

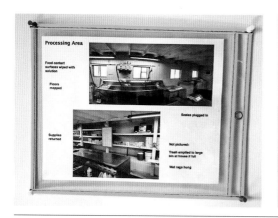

Images posted at eye level quickly communicate standards to everyone on the farm.

- Post photos. This shows workers spaces in their cleaned state and communicates much more efficiently what to do than a checklist.
- Label your spaces. Clear signs—numbers or letters on greenhouses, "Processing Room," "Red-Tag Room," "Freezer Room"—remove confusion for new workers.
- Post whiteboards. Magnetic whiteboards can indicate when 5S tasks should be done and whether tasks have been completed or not.

Before we'd heard about 5S, we never used visual systems. Now we love them. Several years ago, to get a space cleaned, we would tell workers what needed to happen, step by step, walking around with them to point out special instructions, then checking back in once they were almost done. After a while, we began posting daily cleaning checklists in our spaces so we wouldn't have to repeat ourselves so often. The checklists are a great way to tell what needs to happen, but workers often don't read them. Now we just post a large, unambiguous picture of what the cleaned space should look like, with a few special instructions. Workers simply look at the picture and replicate what they see.

5. *Shitsuke:* Sustain

Setting up organization systems is one thing, but integrating them into your farm's culture is another. Sustaining 5S is about applying self-discipline through regular audits.

In some lean factories, a worker is assigned at the end of each week to go around the facility with a clipboard and give a numerical rating based on

5S Is About Feeling Good

Ohno says that while audits are important, the key to making sure 5S happens is proper motivation. The goal is not aesthetics, to make your spaces pretty. The goal, rather, is to improve your work environment so that "people can feel good working there." According to Ohno, cleaning for cleaning's sake "can just use up a lot of paint."[2]

In fact, overcleaning can be a form of waste. 5S is about removing waste and improving workflow so people have a good place to focus and work. Sorting, setting in order, and shining should be used in the service of those goals. Once work is flowing at an optimum pace, it's time to put the paintbrush away.

◼ ◼ ◼

Mihaly Csikszentmihalyi, a psychologist and a pioneer in the study of happiness, discovered through research that people report they are happiest when they are in a state of consciousness called *flow*. Psychological flow, not to be confused with lean flow, occurs when people are completely absorbed in a voluntary activity that stretches their abilities and engages their creativity.

A characteristic of this type of flow is complete focus through lack of distraction. One example is downhill skiing. People enjoy the sport because it requires complete attention. Even the slightest distraction can cause a fall and possible serious injury. Artists report flow conditions when they are deep in their work and removed from all other distractions.

Another element of flow is *order* in one's consciousness. For example, when you are absorbed in a long board game or a game of chess, your thoughts are guided by a clean set of orderly rules, and you are more likely to experience flow. Constraints, like the boundaries and rules in board games or sports, give us order, which increases awareness and focus. As a result, we lose self-consciousness and experience happiness.

Work environments with clear goals, no distractions, and a sense of order—all conditions of leaned-up workspaces—can achieve the same effect.

the cleanliness of workspaces. The goal is a higher rating each week, and rewards are given for the cleanest spaces. Even if you cannot rate your workstations, someone on the farm should be in charge of holding the crew accountable for keeping spaces clean and clutter-free. This can take the form of a daily walk by the farmer, or it can be assigned to a worker. Quick and regular audits will keep waste from creeping back in.

We bundle cleaning tasks with other work assignments to make sure they happen. For example, a worker might receive a to-do list that looks like this:

Harvest 40 bunches of radishes
Process 100 bags of carrots
Clean the processing room

The worker knows exactly what a clean processing room should look like because of the pictures we've posted.

Some time, try having a visitor perform an audit. Anyone should be able to tell the difference between normal and abnormal based on your visual cues, like pictures on the wall or shadow boards for tools. After a jaunt around your farm, the visitor should be able to give you a list of what is missing and which spaces are not up to snuff.

■ ■ ■

In summary, 5S cleans out a farm's arteries. It sets the stage for production. Once daunting cleaning tasks become manageable. A chaotic room in the barn, strewn with old chicken waterers, broken hoses, and dusty feed sacks, might have looked overwhelming before we knew of 5S. Now we have a method to apply. And short, high-frequency straightening and shining mean that we have fewer messes to deal with in the first place.

After a season of cleaning up, we were ready to take waste elimination to the next level. 5S took care of a lot of surface waste. But we knew that wasn't all lean had to offer. As we cleaned, we were learning about tools to root out deeper levels of waste in actual production and management. Our farm felt like a clean slate, and we were eager to start applying these new ideas.

The first step was to learn how to see the opposite of waste, which is value.

CHAPTER 2

Farm for Your Customers: Precisely Identify Value

In 2004 Yuji Yokoya, an engineer for Toyota, was assigned the task of redesigning the Sienna minivan for North American customers. He agreed to the task with one condition: that he be allowed to drive a minivan in all fifty states, in all thirteen provinces and territories in Canada, and in all parts of Mexico.

The result? Yokoya noticed that people drove longer distances in North America than they did in Japan, so they got thirsty and hungry on their drives. Toyota's new Sienna featured fourteen cup and bottle holders and a flip-up tray (big enough to set a hamburger and fries) accessible by the driver. He noticed in Santa Fe, New Mexico, that it was difficult to turn corners on the narrow streets, so in the new Sienna the turning radius was tightened by 3 feet.

Another development was a larger interior, conceived of when Akihiko Saito, in charge of R&D for Toyota, went on a research trip to a Home Depot in Ann Arbor, Michigan. Saito didn't go into the store. He stood in the parking lot. He watched customers get in and out of their cars and load up on lumber, tools, and other items to feed the American do-it-yourself obsession. He approved the Sienna's larger interior once he saw customers frustrated because they couldn't fit 4′ × 8′ sheets of plywood into their minivans.[1]

The Toyota saga is full of stories of this sort: engineers and managers working assiduously to precisely identify customer value.

Many businesses take a glance at their customers and guess what they want. Or they rely on gut feelings about what the marketplace will support. But what customers really want is usually not obvious. It's a mystery to be solved.

▪ ▪ ▪

In the food business, for instance, customers often won't provide straight answers because they fear offense. They keep their real feelings hidden.

Food culture in this country is polite: we tell our hosts, whether in homes or in restaurants, that we liked everything, even if we barely got it down. Everyone wants farm-fresh food, but few are willing to tell farmers what they really thought about their food.

Lean businesses don't rely on hunches. At Toyota, engineers aren't guessing. They're doing more than reading market reports and studying consumer data. They are getting to know the precise habits and needs of their customers and then designing their products—and their factories and work flow—based exactly on what their customers value.

The principle of *genchi genbutsu,* or "go and see for yourself to gain understanding," lies at the heart of lean. *Genchi genbutsu* says that the best way to deeply understand a situation is through personal observation. There is no shortcut to observing actual customer habits as a way to identify value. See for yourself what your customers are cooking and eating and what they are passing by or throwing out.

Precise identification of value is central to a lean enterprise because if you don't know what your customers value, you won't know which of your activities are creating that value and which are contributing to waste. A lean farmer needs a clear understanding of both. If you are wrong about what your customers want, the result is always waste. You won't be able to retain customers and will waste time every year recruiting new ones. Your farm will be rife with overproduction waste because you grew or raised food nobody wanted. In contrast, the more precise you are in identifying value, the more customers you will have and the more loyal those customers will be.

As soon as you find out what customers truly value, this information should guide every decision you make. Don't file the information away. Use it to steer your production and enact changes, or what Womack and Jones call "transforming actions," on your product and in your farming—and the sooner the better.[2] This demand-response loop is the essence of value creation and the key to farming well for both you and your customers.

With Our Own Eyes

When Brenneman asked us where we sold our food, we answered quickly: at a farmers' market, through our farm-based CSA, and to several wholesale buyers—restaurants and a grocery store. But when he asked us what exactly those customers, our community of eaters, valued, we struggled a bit more to answer. We knew all of our customers wanted fresh food. But beyond that? In the months that followed, we set out to answer that question.

Lean requires farmers to discover precisely which foods and services their communities want most.

In Japan, Toyota utilizes its sales crew, which receives the same basic training as all other workers, not just on the car lot but in the design lab as well. In the United States, a car salesperson and an automobile engineer are at opposite ends of a timeline—engineers develop concepts, line workers turn concepts into real cars, and salespeople push the cars onto the public. Toyota closed the loop. Car sellers see buyers every day. They watch them test-drive cars; they talk to them and soak in volumes of information: what colors they like, how soft they want their seats, where they want their cup holders, or how nice it would be if there were a place to store sunglasses. Toyota sends their car sellers to their design labs because sellers *know* more about customers than anyone else. Selling becomes an integral step in product development.

What if we applied the same logic on our farm? We started to see our time direct-selling at our farmers' market as time to develop new ideas, not just sell food. If a customer tells me she used our potatoes in a family recipe for shepherd's pie, I'll ask for the recipe. If I hear that a variety of lettuce is not holding well in the refrigerator, I'll make a note for seed-ordering time. If a customer tells me he likes a certain color of pepper, I'll ask him why and what other colors he might appreciate.

Sometimes we hire a staff person to sell at our farmers' market. Our current staff person is Patricia Oakley. We ask Oakley to send us a market report using a Google Drive spreadsheet. She notes what sold and also communicates customer feedback—for example, if customers say they liked a new variety of head lettuce or if carrots sold better in bags than bunches. Her close observations are part of her work.

One of my favorite ways to observe customers and get feedback is to eat in a customer's home. Not only is it fun to eat our food prepared by others, but I can observe how they cook our food (the kind of food they value), how they store it (the level of packaging they value), or how much they use (the amount they value). Likewise, once a year we try to eat in the restaurants we deliver to. I talk to the chef while I'm there, but I also watch. How is our food presented on the plate? How is it described in the menu? What other food items on the menu might we be able to supply?

Friends on Facebook sometimes post pictures or recipes that used our food. This is another opportunity to take notes and observe how customers use our products.

Most farmers started out farming because they had a passion for raising animals or growing grains or vegetables or fruit. Perhaps it motivated them

Farming for your customers means listening to what customers value most and building that value into your products. A clean presentation can be part of value.

to quit more stable jobs and invest their lives in their passion. Most farmers I know have a genuine fascination with what they are producing. The same amount of fascination we have for our crops or our animals should be applied to our customers. This means researching what customers value with the same zeal and diligence we use to research our production.

Avoiding Distortions:
The Customer Alone Defines Value

This principle seems obvious, but over and over again farmers let other factors define value and drive their farming. We have been guilty of each of the four distortions below.

1. *Technology fascination.* Many farms get bogged down with tractors that are loaded with too many features, greenhouses that are too complex, or overly expensive animal feed systems, to name just a few examples. Even though farm technology improves every year, invest in it only when it will obviously create more value at lower cost for the customer. On our own farm, I have purchased several seeders that fascinated me but did not match our needs, as well as electronic irrigation timers that were much more precise than we really required.

   Womack and Jones argue that German industry declined after the Cold War because engineers, not customers, defined value and drove business decisions. "Designs with more complexity produced with ever more complex machinery were asserted to be just what the customer wanted. But where was the evidence?" They recall conversations with German firms in which product failures were "often explained away as instances where 'the customers weren't sophisticated enough to grasp the merits of the product.'"[3]

2. *Product fascination.* Don't assume customers will want to eat what *you* want to produce. You might be fascinated by purple beans or multicolored carrots or escarole, but are your customers? In the first few years, we grew too many of these types of specialty crops simply because *we* wanted to grow them.

3. *Process fascination.* Whether it's growing on rooftops or the latest new hydroponic system or vertical growing systems, many farmers let a fascination with process obscure what it is their customers really want. Not that new ideas don't have a place—trying new things is an important lean principle—but we go wrong when new ideas, not the actual wants of customers, drive our farms.

More Tips for Finding Value

A full picture of the value your farm creates can take years to develop and might be more complex than you realize. There are many elements of value.

Customers Almost Always Value a Mix of Goods and Services

Many farmers assume that what customers value is a *product*. In reality, in the food industry, this is rarely the case. Customers almost invariably value some combination of goods and services.

One of our restaurant customers orders a predictable amount of garlic every year—about 3,000 heads. We grow a hardneck variety with large cloves that works well in a roasted garlic dish on her menu. The customer no doubt values our product. But she would not order from us if the garlic didn't come with service components. For example, the chef prefers to receive the garlic in small quantities throughout the season rather than in one large delivery. So consistent delivery is part of value. She wants to slice the top of the garlic and roast it whole with as little prep as possible so cleaning and selection (for large cloves) is part of value. She wants to put fresh garlic on the menu as early in the spring as possible, so a phone call to let her know when it's ready is part of value.

There Can Be Different Value Streams on the Same Farm

Brenneman told me the last time I visited the Aluminum Trailer Company that they are careful not to assume all of their customers value the same things. They have customers who want the most economical trailer possible, but they also have high-end customers who want specialized trailers with a high degree of customization. These latter customers want a higher proportion of service in addition to a quality product.

Likewise, if you raise beef cattle, the auction house is likely looking for a different product and service mix than your retail customers. If you raise animals for milk, the wholesale distributor probably

4. *Letting supply determine value.* Despite careful planning, all farms fluctuate from time to time between oversupply and undersupply. The amount of inventory you have, however, is irrelevant to customers. It doesn't change what they want or the amount they want it in.

One concrete example is kale in CSA boxes. A common reason people leave CSA programs is that their farmers gave them too much kale. These growers might have valued getting rid of an oversupply, but their customers didn't value receiving it.

Similarly, Womack and Jones assert that today's airline industry, dominated by large jets and layovers, illustrates a similar habit. Airline comes with a different set of values than herd share customers. It can help to separate customers into groups and identify what each group most wants.

Sometimes the Farmer Knows Best

Apple's founder Steve Jobs famously eschewed customer research. He said that a lot of times people don't know what they want until you show it to them. The same is sometimes true with farm-direct food. Many items we grow don't show up on grocery store shelves. The products are unique. Only a few customers have even heard of some vegetables we grow. But I know that once customers see and taste them, many won't resist.

When we decide what to pack in CSA boxes, we envision a box filled with a wide range of fresh food from our fields. We try to grow what customers will eat, but only *I* know what's at peak quality on our farm any given week. For unusual items like tatsoi or Tokyo bekana cabbage, I have a sense for what our customers will want and how much

they'll use, even if they've never tried it. Over time we have realized that it's best to include these types of unusual items no more than once per week.

Customer Values Change

Carmakers redesign their models every four to five years to adapt to changing customer values: better fuel economy, more interior room, less weight, and so on. Food producers should do the same. New and better fruit and vegetable varieties are developed every year. The way people eat and what they eat changes. Don't get too stuck on a set formula.

Even if demand for your food product itself doesn't change, delivery mechanisms are in constant flux. For example, many farm-direct producers are finding out that compared to ten years ago fewer customers today are willing to travel to farms to pick up their food. These customers are shifting what they value: they still want farm food but now prefer more convenient delivery. This is one reason lean is an ongoing process of continuous improvement rather than a linear process with a defined end.

customers want "to get from where we are to where we want to be safely with the least hassle at a reasonable price." Whereas the airline's definition of value "seems to involve using their existing assets in the most 'efficient' manner, even if we have to visit Timbuktu to get anywhere."[4]

Clay Bottom Farm's Customers

When Brenneman first visited our farm and introduced us to the concept of value, we ticked off ideas that came to our heads about what our customers valued from us. The real work, however, lay ahead, as we sent out online surveys, asked direct questions, observed in restaurants, and ate in a few customers' homes for dinner. Below I list our conclusions about what our customers want from us.

Your customers may not value the same things, and you may not want to respond the same way we did. But we can say from experience that if you move customers to the center of your farm, you will have happier and longer-lasting relationships with them.

CSA CUSTOMERS

Meal prep help. Our CSA customers are busy. They appreciate the *service* aspect of the CSA—of someone else assembling a portion of their weekly groceries. We are taking that service a step further by building CSA boxes around a meal plan. We advertise our CSA as a meal in a box, and we include recipes with each week's newsletter. Customers don't want an eclectic mix of food they don't know what to do with. They want meal help.

A farm connection. Many customers bought our food for more than the food. They also wanted to be part of a farm. "You need to tell your story," one customer told us. In an effort to do just that, we started posting two times per week to Facebook. We ramped up our weekly CSA newsletter (pictures of people and cute babies help). We hosted meals and open houses at our farm.

For a group of customers who live an hour away, we organized a pickup rotation so that one person from the area comes to pick up everyone's box from that area. The move saves us from driving and connects customers to their food. "We love the rotation," one customer told us. "It gives us a reason to come see the farm."

Flexible pickup. Customers with busy schedules can't always pick up their boxes at *our* convenience. We used to give customers a window of just a few hours to pick up. Now we've installed refrigeration units and a number-coded lock box at our town pickup location so that customers can pick up their boxes at any time.

The right mix and the right amount (not too much). Over the years we've tried to hone the balance of what we offer. Greens are our specialty, and many people sign up just for salads, so we offer greens in every box. For other items, we make sure to include enough for a dish. The general rule is fewer items in each box—never more than ten—and no tiny portions of a single item.

With total box volume, the trick is not giving too much. Our customers told us over and over again that they hate to waste food. So we

We aim for the right variety and the right amount in our CSA boxes.

are very careful to make sure the boxes are not overstuffed, even if there is a lot of food on the farm. In addition, we started taking a week off every two months to let customers clean out their refrigerators. This gives us a break as well and schedules in time for us to travel during the summer—a luxury we rarely enjoyed in our early years of farming.

FARMERS' MARKET CUSTOMERS

Blocks of color. Our primeval selves come alive in shopping for food. When our ancestors gathered forest food like berries to survive, they had to be attuned to notice blocks of color. We observed the same habit in farmers' market customers. Color attracts. Whenever we can, we bunch similar colors together for an attractive display. And we *show* color, for instance by turning radishes upside down so that their red roots face up instead of down. Tall, abundant stacks sell better than a scanty display. As the saying goes, "Stack it high and watch it fly."

Wide selection. We noticed that market customers prefer standard varieties (red tomatoes more than heirlooms, for example) and large-sized vegetables. Still, they appreciate a wide, even eclectic selection. It's part of the appeal of going to market. So we aim for variety and don't shy away

Calling a chef to tell her what is available on the farm is part of our product rather than an inconvenience.

from uncommon items like turmeric or figs every so often, even if only as a way to draw interest in our other items.

WHOLESALE CUSTOMERS

Timely delivery, two days per week. Our chefs appreciate wholesale deliveries on Tuesdays to load up for the week and on Fridays to gear up for the weekend. They told us that they want us to get there early—before 4 p.m., so they have food on hand when dinner customers start showing up at 5.

Timely communication. Wholesale customers told us they want a heads-up when new items are available. If spring asparagus will be ready on May 15, we try to tell them a week ahead of time. In fact, every week we make a phone call or send them a text message (their choice) to let them know what's ripening on the farm. This allows them to put in an order while flipping eggs or plating a dish. Rather than an inconvenience, we consider the call or text part of our product.

Income and Food Value

What customers value in food can depend a lot on their economic circumstances. (As ever, there are plenty of exceptions to these rules.)

Many lower-income customers want a good *value*—they want as much food, as many calories, as their dollars can stretch. They also want familiar food that they know how to prepare. And they value receiving food in their local neighborhoods. To serve these customers, cut costs to keep prices affordable and deliver to neighborhood markets close to their homes.

The primary focus for many middle-income customers is *quality* food. These customers care more about flavor than fancy packaging or even the cheapest price. In addition, they value food information to help them determine which foods pack the most nutrition and taste. To serve these customers well, spend time sorting out seconds and develop ways to communicate facts about your food.

A main concern for many wealthy customers, in addition to quality, is *presentation*. These customers want first and foremost clean produce, attractive packaging, and service with a smile. To serve them well, ramp up the service aspect of your farming—hire a logo and web designer, buy labels, and do your best to make sure no item leaves the farm dirty. Don't be afraid to sell at a higher price—service has a cost.

These lists aren't exhaustive—all of our customers, for instance, value quality and freshness. And we discover new value attributes every year. The lists do represent values that were not always obvious to us until we started really looking with our own eyes.

Finding Value from Other Farms

I've seen astute awareness of customer values on other farms. A good example is at Lehman's Orchard. Owner Steve Lecklider told me that for a long while he had suspected customers were interested in his products—which include value-added goods like wines, cider, vinegars, and dried fruit—year-round, not just in the summer. But sales at his on-farm retail store always slumped in the winter. "Then I realized that the problem wasn't necessarily what I was offering but location," he explained. "I learned my customers don't want the hazard of driving down a rural Michigan road in the winter to do their shopping."

So when a warehouse opened up in nearby Buchanan, he approached city planners with an offer to turn it into a downtown retail space for his goods. "I'm taking my products exactly where the customers want them," he told me. If sales go well, he plans to add a winery and microbrewery—moving current value-adding processes from his farm to his new retail location. This will allow customers to see more of the process and still feel attached to the farm.

On a larger scale is Van Belle Nursery in Abbotsford, British Columbia. The company has 80 acres and fifty year-round employees (100 in peak season) and grows vegetable and fruit starts as well as potted plants for growers across Canada and the United States. DeVonne Friesen, vice president of business development, said the company noticed that young people weren't getting into gardening because of a knowledge gap: they don't know how to "water, prune, fertilize, and overwinter plants," she explained. In fact, according to the company's research, only 2 percent of consumers are master gardeners. Most are "casual or reluctant" gardeners.

Consequently, the nursery worked to develop varieties that are easier for beginners to grow— for example, plants that are more disease-resistant or need less pruning. "We always ask how an improvement makes life easier for the person buying the plant," said Friesen. To this end, they also created a new mobile-friendly website that explains in plain language to a new generation of growers which plants work best in specific geographic areas and basic care instructions.[5]

At Sustainable Greens, the Three Rivers, Michigan, farm where Rachel and I worked for several years honing our skills, awareness of what customers valued was paramount. James Lind, as field manager, planted and tended crops with his customers' values in mind. I remember being given meticulous instruction on how to cut greens and handle leaves without bruising them, as per the wishes of chefs receiving the greens. Likewise, his wife, Kate, made sure their produce was washed and packaged just as the chefs ordered.

One such chef was Grant Achatz, who grew up flipping eggs and frying bacon as a short-order cook in his parents' working-class Michigan restaurants. After graduating from the Culinary Institute of America in New York, he went on to work for Thomas Keller in the prestigious French Laundry restaurant in Yountville, California. In 2005 Achatz founded his own restaurant, Alinea, in Chicago. A year later *Gourmet* magazine called Alinea the "best restaurant in America." In 2011 Alinea was voted the sixth-best restaurant in the world and the best in North America by *Restaurant Magazine*, and it was the only restaurant in Chicago awarded three stars—the highest rating possible—in the 2012 Michelin Guide.

Needless to say, Achatz had standards. And the Linds listened. They won his business because they gave him what he valued. Every Wednesday night Kate brought Achatz the highest-quality microgreens, petite carrots, arugula, herbs, salad greens, and rare forest finds. But besides food, she also provided him with stories of the farm. Achatz valued both. "Kate was part of our family," he wrote in his autobiography. Kate "talks to chefs directly and tells them what looks good on the farm a couple of days prior to delivery. . . . I would constantly bug her for obscure ingredients that I had read about in old cookbooks, and she would promise to plant or forage for them if they weren't readily available."[6]

Once, at Trio, a Chicago restaurant where Achatz worked prior to Alinea, he wanted to prepare a special dish for William Rice, an influential national food writer, and David Shaw, chairman of the James Beard restaurant committee. The dish featured angelica, which he had asked the Linds to grow for him. When Kate brought in the delivery, she apologized and told Achatz that her husband had accidentally tilled in most of the angelica; only two stalks remained. Achatz wanted them anyway. He intended to clean the branches and let his guests slurp their food through the hollow stems. The Linds' story unexpectedly became part of the dish:

> I began removing the [angelica] from the jug with the intent of snipping away the leaves and paring them down to a single straw, when I stopped. They looked like flowers in a vase, they were alive, and they were a part of that small farm in Michigan. We needed to serve them that way. In fact, that needed to be the entire point of the course. . . . I wanted Bill and David to remove the branches from the glass vase that we would serve them in, contemplate the angelica, its history in gastronomy, and hear about Kate and her tiny farm five hours away. What I put inside for them to eat was almost irrelevant.[7]

According to Achatz, the dinner "went wildly well." Shaw wrote an article for the *Los Angeles Times* about the experience "that made its way to all the food forums online and likely the desk of every writer who cared about what was happening in the food world."[8]

It's hard to provide a customer more value than that.

As these examples show, customer value is foundational, and a right sense of value will elevate your farm. Once you have a handle on value, the next step is to root out waste as you produce and deliver that value. With a better sense for what our customers valued, Rachel and I were eager to start this process. Learning to *see* value, not just in the hands of our customers but as it skips and hops around our farm, was next.

CHAPTER 3

Learn to See Value

Take a minute to imagine the flow of one carrot on your farm. Strip away everything else and just focus on the carrot, sitting in the ground or suspended in air as it is moved from one place to the next, as it gets more and more valuable.

Start at the very beginning, with your idea to grow the carrot (called *concept*) or perhaps with an actual advance order. What steps are involved? How long does the process take? What equipment (computer software?) are you using, and what is it costing you?

Then move on to ordering seeds. How long does it take? When seeds arrive, where do they go, for how long?

Then imagine the seed going into the ground. What steps did you take to prepare the soil and plant? With what tools? How long did it take? How many people were involved?

Then image the carrot in the ground, getting bigger, gaining in value. What are the costs to tend the crop—hoeing, irrigating, or spraying—to ensure a good result?

Then, through harvesting, washing, cold storing, bunching or packaging, and delivering the carrot, imagine the line the carrot takes across your property and across town, perhaps farther. Was the line smooth or twisting or ragged? Did it double back? How many hands touched the carrot? What motions were required to display the carrot? How long did the harvested carrot wait before being received by its customer, the eater?

Visualize getting paid. How many motions were required, what equipment was needed, and how much time passed before the payment became cash?

Note that along the way, the carrot—from petite, to baby, to full size, and then bunched with other carrots and washed—gets more valuable as it gets closer in time to receipt by the customer. This is called *value creation*.

A value stream map of 275 carrot bunches at our farm.

1. Plan the project (type 1 *muda*, a Japanese term for waste or any activity that does not add value. See page 55 for a description of type 1 and type 2 *muda*). Review sales and customer feedback, 15 min.
2. Order seeds (type 1 *muda*), 10 min., $9.50 for seeds, $5 est. value of overbought seeds (type 2 *muda*)
3. Chisel plow, 10 min. Till, 10 min. Shape with bed rake, 10 min. Apply compost, 20 min., $15 for compost
4. Plant (value adding), 15 min.
5. Cover with row cover (type 1 *muda*), 20 min.
6. Hoe (type 1 *muda*), 20 min. × 2
7. Irrigate (type 1 *muda*), 15 min. to set up timer, $10 est. cost of water
8. Irrigate (type 2 *muda*); water runs in unexpected rain, 15 min., $5 est. cost of water
9. Harvest (value adding), 2 hr.
10. Wash and bunch (value adding), 1 hr., 50 min.
11. Store in cooler (type 1 *muda*), $7 ($1/day for electricity)
12. Deliver (type 1 *muda*): to farmers' market, 3 min. (30 min. divided by 10: carrots = 10% of items), $2 for gas; to restaurants, 3 min., $2 for gas
13. Display (value adding), 1 hr. (1/10 of sales over two 5-hr markets), $5 booth rental for 2′ table space. Information loop: Receive costumer feedback
14. Receive money (type 1 *muda*), 10 min. for accounting
15. Deposit cash (type 1 *muda*): trip to bank, 3 min., $3 for gas

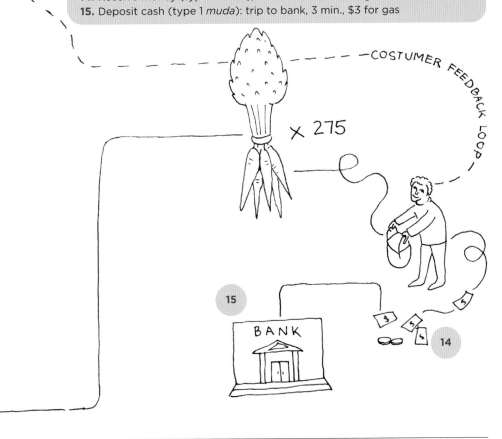

× 275

COSTUMER FEEDBACK LOOP

15

BANK

14

Variable supply costs	=	($63.50)	
(Type 2 *muda* = $10.00)			
Type 1 *muda* time: 2.5 hrs × $15	=	($37.50)	
Type 2 *muda* time: .25 hrs × $15	=	($3.75)	
Value-adding time: 5.1 hrs × $15	=	($76.50)	
Total costs	=	($181.25)	
Sales: 275 x $2 (avg. price)	=	$550.00	
Profit (without fixed costs)	=	$368.75	

Totals From Value Stream Map.

Compressing Steps in the Value Chain

Remember Ohno's goal, to reduce the timeline between concept and cash? With the carrot, you've just imagined the timeline. Now let's shorten it. Imagine a customer ordered the carrot from you with lightning efficiency, perhaps through online order-receiving software through which the customer has prepaid. At seed-ordering time, you choose a company with overnight delivery, and you order the day before you seed, so the seed goes straight from mailbox to ground. You've developed quick systems for bed preparation and seeding, so the process takes just a few minutes.

Since you've used plenty of compost and set up irrigation timers, the carrot grows quickly. Early one morning, when the carrot is at peak freshness, you harvest, wash, and bag with just a few motions. Your cooler is

Clay Bottom Farm carrots. Value stream maps help us locate waste in our carrot production.

close by, so in a few more steps the carrot is safely stored. At delivery time, you back up a vehicle to the cooler door, load up the carrot along with a truckload of other items, and drive off. The carrot has traveled in straight lines and has gained all of its value with only a minimum of interruptions.

Once we knew what our customers valued from us, the next task was to visualize our products as they traveled around on our farm, gaining in value. And then we tried to compress the steps.

Brenneman helped us out. In our processing room one afternoon, he pulled out a large sheet of paper, sticky notes, and markers. We spent a good portion of that afternoon mapping out value creation for two of our items—a 50-pound box of tomatoes and a 3-pound bag of salad mix. Sticky notes represented steps in the value chain. On each note we wrote the activity involved, the time each step took, and all costs involved in production (*variable costs*).

The idea was to learn how to *see* value, to distinguish value creation from wastes, like waiting time and unnecessary movement. The exercise helped us calculate our costs, figure out our hourly wage, and start to identify targets for getting faster. This work—learning to see value—is the first and primary task of lean managers.[1] The work requires honesty—naming reality for what it is so that later you can discern waste among true value-adding activities along the entire value stream.

Farmers with lean instincts and good imaginations are constantly doing this work in their heads. They know which steps are adding value, and they focus their efforts there. For many, however, the task does not come so easily. It's often not obvious how value is created, where it's created, or when it's created, especially when a lot of waste muddies the view. It sometimes helps to lay out the steps on paper and collect data with each step. Other times we accomplish the task more efficiently with our eyes and our imaginations.

Value Stream Maps

When value creation is spelled out on paper, the exercise is called *value stream mapping*. The practice is commonplace in lean industries, where maps are often large, covering an entire wall, with sticky notes denoting stops along an assembly line or stages in a manufacturing process. The maps can get complicated quickly. Larger firms that consistently use value stream mapping often have software programs designed for the purpose.

For example, managers at a plant that assembles bicycles might create a map by starting with a sticky note labeled "parts receiving" and ending with

Tips for Using Value Stream Maps on Your Farm

Value stream maps aren't always necessary, but they're helpful. For most farms, simple maps with basic information are enough to give farmers a bird's-eye view of how value flows in their operations. Even a small diagram using boxes and arrows can yield valuable insight. Larger farms would use more complex mapping. The steps are simple, don't take a long time, and let you see your business from a new angle.

Start with a Problem You Are Having

Value stream maps are a great way to problem-solve. Addressing a specific problem can help focus your mapping. For example, if your labor costs are too high, use a value stream map to track each step in your process and record data on labor used in each step. Value stream maps can also help reduce transportation waste and help identify and solve production bottlenecks. Pejstrup, the farm consultant from Denmark, told me that value stream maps help her and the farmers she works with sort out these kinds of process kinks.

Look at Services as Well as Products

If your customers value a farm connection, then track the ways they connect with your farm, including the time and cost to you to provide those services, even if the delivery of the service is not as linear as the delivery of your product. Include steps customers must take to connect with the farm, like their initial website search, a call for more information, and farm tours. Remember, the service aspect of your farm is just as important as the product.

Assess Information Flow, Not Just Physical Transformation

Remember the demand-response loop, where you use information about your customers to make changes to your product? Womack and Jones say that loop should be visible on your maps. They suggest tracking "the flow of *information* going back from the customer to the producer with the *transforming actions* on the product."[2]

In the case of our carrot, let's say a customer eats the carrot and provides feedback (positive or negative) to your farm. What is the flow of the customer's voice? How many ears does it pass through? How many databases? At what point does it reach the farmer, if not immediately? Once the voice is heard, what happens next? Who responds?

one called "customer distribution." In the middle might be steps like "tire fitting," "derailleur assembly," and "painting."

Along with each step, map creators record data such as setup time, processing time, units in queue, complexity, percent defect, number of employees required to fabricate or assemble parts, or cost of raw materials. Once the material flow is laid out, the maps may be overlaid with another layer

How quick is the reply? Who determines whether feedback is worthy of causing a change in your product? How does good feedback translate into changes in how future carrots are grown, washed, or delivered?

Determine Your Hourly Wage

Value stream maps are a great way to determine worker productivity per hour. Track your time and your costs for each step in the process, then divide by sales for the same number of units. This kind of cost accounting can be done in line fashion on paper, but the value stream map gives you a better visual idea of where you are spending your time.

Include *All* the Steps

It might seem obvious, but often farmers forget steps, intentionally or not. For example, the time you spend at the livestock auction is a step. Your time unloading the delivery van is a step. Your time entering invoices is a step. I suspect that many of the most frequently forgotten steps are ones that don't add value.

In particular, farmers should not forget to track steps at the very beginning and at the very end of a product's lifespan. At the beginning is concept work—time spent on planning and research, deciding where to plant a crop,

and what method to use. Like any other part of the value stream, this work should be leaned up. At the other end is selling. The time you spend on financial transactions like invoicing and banking are legitimate targets for leaning as well.

Create Both Current State and Future State Maps

The current state map is a snapshot of your production process now. The future state map details your lean vision. Which forms of waste will you target first? What changes will you make? How will that change the value stream? Once you know current reality, map out a vision for future work.

Look at the Whole Picture (Relationships between the Steps)

One of the benefits of the value stream map over a simple list of steps is that you can see relationships between steps. Sometimes entire steps can be removed. At other times, you might notice a lot of transportation or friction from one step to the next. Look at your sticky notes— your steps—then look at the gap between the notes and ask, "How long did it take our product to jump from one stage to the next? Were there costs? How could the product flow more smoothly?"

detailing information flow: "order receiving," "initial response to customer," "order relay to assembly line," "parts for production ordering," and so on.

These aren't simple flow charts. By the end, value stream maps can have crisscrossing lines going in many directions, resembling complex board games. Flow charts detail where a product moves. But to lean up, you need more information than *where* your product has been. You need to know

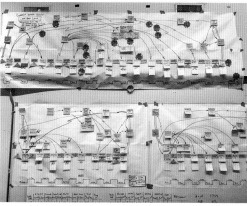

Value stream maps at Bjornemosegaard I/S dairy farm in Faaborg, Denmark (left), and Aluminum Trailer Company in Nappanee, Indiana (right). Value stream mappers often use sticky notes to track production steps. The maps give producers a bird's-eye view of their operations and make it easier to spot waste. Left photo by Susanne Pejstrup, Lean Farming®; right photo courtesy of Steve Brenneman, Aluminum Trailer Company.

costs and you need to analyze the relationships between steps to get clear pictures of which steps lead to value creation and which are full of waste.

After a 5S decluttering and working to name and see value, we were ready to put on our gloves and get to work stripping away everything that didn't add value—our waste. Though we had literally removed tons of physical waste, we knew there was still a lot of process waste. Our next task was to learn to see waste, just like we learned to see value. This is the subject of the next chapter.

CHAPTER 4

Ten Types of Farm Waste

*It is astonishing as well as sad, how many trivial affairs
even the wisest thinks he must attend to in a day.*

—HENRY DAVID THOREAU

When we think of waste, images of trash cans full of plastic or rotting vegetables come to mind. But in lean, waste involves more than that. The word for waste in Japanese is *muda*. It encompasses concepts like idleness, futility, and uselessness, in addition to physical waste. In lean, it means any activity that does not add value.

In fact, according to lean thinking, only three types of activities can ever occur on your farm:

1. Actions that add value
2. Actions that do not add value but are necessary (type 1 *muda*)
3. Pure waste (type 2 *muda*)

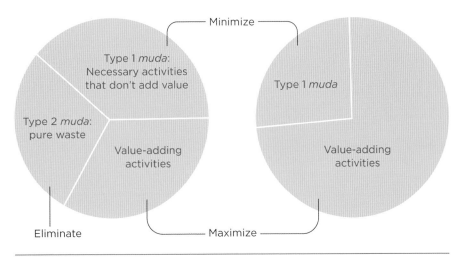

3 Types of Activity on a Farm

Your goal is to move as many of your activities into the first category as possible and then to perform those activities as efficiently as possible. Type 1 *muda* should be kept to an absolute minimum, and type 2 *muda* should be banished altogether as soon as possible.

Examples of actions that add value might be potting out transplants or washing your food. The farmer is not alone in creating value. When cows eat grass, value is added. When plants turn sun into food, value is added. In many ways, the function of farming is to set the stage for the sun and plants and animals to do the real creating of value. These direct actions on the end product make the product more desirable and more valuable.

Examples of type 1 *muda*—the necessary actions that don't add value—might be keeping meat frozen as it awaits paying customers or storing grains in bins or cultivating a bed of spinach. Whenever you set up irrigation, vent greenhouses, or move portable fences, you might be performing an important—even necessary—task. You might be setting the stage for nature to perform its value-adding magic. But you yourself are not adding a bit of direct value to your product. Lean says strive to minimize or eliminate these actions.

Examples of type 2 *muda*—pure waste—might be letting milk become contaminated, leaving crops in the field, or packaging more than necessary. When you grow a crop nobody wants, order too many seeds, or let cut hay mold in the field, you are adding waste to your farm. These actions are often easiest to see (or smell, in the case of rotten food). Because they add no value and only add to your costs, they should be eliminated.

All of your activities fit into one of these categories. These are tight definitions. There is no fourth category. Since the focus of a lean enterprise is waste elimination, lean managers spend a lot of time analyzing their work and categorizing their activities into these three types.

A problem with many businesses is that they think leaning up simply means getting rid of the obvious waste—type 2 *muda*. "If we get rid of unneeded tools, grow the right amount, don't waste a single seed, doesn't that make us lean?" a farmer might ask. No, it doesn't. Type 1 *muda*—necessary tasks that lead to value but don't yield actual value—are actually more pernicious and prevalent. As Shigeo Shingo, a Japanese engineer many call the world's leading expert on manufacturing practices, observed, it's only the last turn of a bolt that tightens it—the rest is just movement. Type 1 *muda* on the farm—the time you spend plotting out the garden or ordering seeds or moving a hoe—is just movement. It does not directly add value. These activities should receive the same scrupulous attention for improvement or, better yet, elimination as every other activity.

This concept of *muda* versus value adding is sometimes hard to grasp, so let's look specifically at the example of the carrot. Remember the carrot's journey from Chapter 3? Here's a list of the steps in that journey:

- plan the project
- order seeds
- prepare soil
- plant
- hoe
- irrigate
- spray to ward off pests
- harvest
- wash
- store in cooler
- bunch or package
- deliver
- display
- receive money
- deposit cash

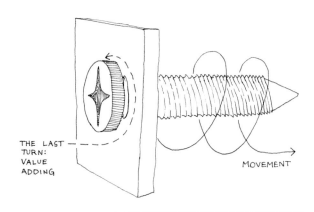

THE LAST
TURN:
VALUE
ADDING

MOVEMENT

Shingo's bolt. Shigeo Shingo noted that the last turn of a bolt tightens it—the rest is just movement. Lean helps farmers analyze their work to discern which actions truly add value.

The question for a lean farm manager is, which motions are direct actions on the object causing its dollar value to go up? Those are the actions that add value. Here's the list again, but this time I've added categories according to the three types of activity:

- plan the project: type 1 *muda*
- order seeds: type 1 *muda*
- prepare soil: type 1 *muda*
- plant: value adding
- hoe: type 1 or type 2 *muda*
- irrigate: type 1 or type 2 *muda*
- spray to ward off pests: type 1 or type 2 *muda*
- harvest: value adding
- wash: value adding
- store in cooler: type 1 *muda*
- bunch or package: value adding
- deliver: type 1 *muda*
- display: value adding
- receive money: type 1 *muda*
- deposit cash: type 1 *muda*

Notice that many of the steps needed to produce the perfect carrot and turn it into cash were necessary but did not directly add value—type 1 *muda*. When you called the seed company to order the seed, you did not add value to the seed. When you moved your seed from the mailbox to your storeroom, you did not add value. When you charted out your field maps and planned your growing strategy, you did not add value. These activities move the bolt but don't tighten it.

A few activities—hoeing, irrigating, spraying—could be type 1 or type 2 *muda* depending on whether or not they were truly needed. Either way, when you weeded, the actual carrot did not become any more valuable than it was before you weeded. So weeding at best should be kept to a minimum.

By contrast, you *did* add value to the seed once you put it in the ground. Seeding is a direct action on an object that makes it more valuable, as long as conditions will allow it to grow.

Then nature takes over. Within a few weeks, the nascent seed is now a tiny hair of a carrot. In the right market, someone will pay for pencil-thin carrots (we've sold them on occasion to a high-end restaurant). A few weeks later, the carrot is petite size and worth a few cents more (again, most likely in a specialty market). Eventually the carrot reaches full size. If it is very large and good quality, it's worth perhaps twenty or thirty cents to someone willing to come out to your farm and harvest the carrot.

Pull the carrot and wash it, and you've added much more value. Now a customer might be willing to pay you thirty or forty cents. Bunch or package that carrot and you've increased the value further still. Now it might be worth fifty cents. Display the carrot at market—pull it out of a bulk tote and show off its orange color—and you've taken another direct action that added a few cents of value.

Notice that most of the value created happens at seeding time and during harvest and post-harvest activities. These are the moves that tighten Shingo's bolt. Tending plants and animals is the fun part of farming but not always where the most value is created. We've noticed on our farm that as we lean up, we spend more and more of our time either putting seeds in the ground or in harvest and post-harvest activities. In part this is because we've become efficient at growing, but it's also because that's where the value is. We are making money when the seeder is drilling lettuce and when we are pulling tomatoes off the vine. Just about everything else is type 1 or type 2 waste—to be eliminated or minimized. That's why we've leaned up our bed preparation, automated our irrigation systems, installed automatic vents in the greenhouses, and kept weeding minimal.

The carrot story also illustrates how farming is an intimate partnership with nature. Many tasks that add value are really joint ventures in which the farmer creates an ideal condition for nature, through the power of the sun, to do the real work of adding value. Farmer and writer Joel Salatin calls this "farming the sun." Farmers do not create value on their own. This is a key area in which farming diverges from manufacturing. In factories, humans weld steel

Harvesting and washing peppers and seeding ginger are examples of direct actions that add value to products. Pepper photo courtesy of David Johnson Photography.

Trellising tomatoes and field prep are examples of type 1 *muda*, actions that might lead up to value but that do not directly add value. Lean says to minimize them.

to steel in order to add value; on farms, humans and nature cocreate value. This means a farmer can't ignore the impacts of his or her actions on the land and with animals. Farmers who do so risk overreaching with lean (see Chapter 12).

Ten Types of Waste on Farms

Taiichi Ohno identified seven types of waste in Toyota factories. I list these original seven *muda* below because they are ubiquitous on farms as well as in factories. To them, I add three more concepts of waste particularly common on farms.

1. Overproduction
2. Waiting
3. Transportation
4. Overprocessing
5. Inventory
6. Motion
7. Making defective products
8. Overburdening (*muri*)
9. Uneven production and sales (*mura*)
10. Unused talent

1. OVERPRODUCTION

In farming, overproduction in the form of unsold crops or animals is among the most odious kinds of waste, because unsold goods have a lot of investment wrapped up in them and often cost money to get rid of. Overproduction can happen because of poor planning (erroneous forecasting), a bumper harvest (unpredictable weather), or market volatility.

I include in the definition of overproduction waste the practice of selling items at lower prices to clear out excess inventory or oversupply. The energy you exert to sell and manage those crops is time and energy you could have spent producing items that customers place more value in. Displaced energy is wasted work.

2. WAITING

On the production line, waiting waste takes the form of workers standing idle until parts arrive or equipment is fixed. Waiting waste also occurs when a product sits, as when crops or animals that are ready for market await customers.

When people are underutilized, it is obvious how waste is generated: you are paying workers to stand around. When products sit around, the waste is less obvious but still present. Every time you store an item there is a cost—for the building, for conditioning (if needed), for moving the item again later to its next destination, and for the mental space required to remember what you have and where it is.

3. TRANSPORTATION

Moving goods from one place to another happens every day on farms; so does transportation waste—the inefficient or unnecessary transport of products. Examples might be inefficient equipment use—using a tractor to carry a single bunch of carrots or making four hay-loading trips with a small wagon rather than one trip with a big wagon—or delivering products that customers would be willing to pick up at the farm. Many direct market farms get bogged down with poorly planned delivery routes, where farm products are delivered in small batches to far-flung accounts rather than consolidated to minimize road time.

4. OVERPROCESSING

This type of waste encompasses any activity that creates or does more for your customers than they are willing to pay for. Examples include bagging items that could be sold without packaging, washing food more than is necessary, delivering to more locations than necessary, or spending too much money on websites.

5. INVENTORY

Inventory waste means keeping more materials or goods on hand than is absolutely necessary. On farms, inventory management can be challenging because production output is impossible to control completely, since nature always finds a way to alter a farmer's plans. Even the best production forecasting will never allow a farm to determine exact yield, compared to a factory that can make exactly the number of units it needs. Even so, farms can do much to keep inventories—of both supplies and finished goods—to a minimum.

6. MOTION

Too much moving is a form of enormous waste on farms. Motion waste includes handling items too many times, inefficient harvest practices, and poor planning at planting or seeding time (running back to the greenhouse for more trays of seedlings). A common problem on many farms is spreading out too far—propagation greenhouses too far away from fields, fields too far away from processing areas, processing areas too far away from storage rooms, storage rooms too far away from loading docks, loading docks too far away from the road. Awkward farm layout also contributes incredible motion waste, for instance when you have to go around three buildings and cross a road to bring home a harvest rather than make a straight path.

And almost all farms suffer at times from the waste of looking for misplaced tools or from walking too far to retrieve tools stored in faraway locations. The 5S system (see Chapter 1) is designed to specifically target this waste.

7. MAKING DEFECTIVE PRODUCTS

Defect waste includes unsellable food and food that must be discounted because of poor quality. Defects result for many reasons. For animal products, poor management increases animal sickness and mortality. For fruits and vegetables, poor handling, improper storage, and poor field management are among the many reasons crops don't turn out the way farmers intend. Again, because farmers live and work in the messy world of nature, some causes of defects, such as harsh weather or insect migration, are outside of a farmer's control.

Defects are a major source of waste because, as with overproduction waste, defective products often contain a lot of lost investment. It's best to spot defect early. We would much rather a crop fail within a few days of planting time than after we've spent time and money growing and tending or even harvesting a crop. The lean principle of *poka-yoke*, or "mistake-proofing," targets this waste through systems for early defect detection (see Chapter 5).

8. OVERBURDENING (*MURI*)

In the Japanese language, *muri* is often used to mean "impossible," "unsustainable," or "unreasonable." On the farm, *muri* waste occurs when workers and equipment are overstretched. With people, *muri* leads to burnout, injury, and poor work. With equipment, it leads to engine failure, broken handles, and worn-out parts.

Equipment and bodily overburden can be a problem on farms especially around harvest time, when there is more to do than time allows. And there is often a lot of *muri* when farms grow too rapidly. Workers are overstretched trying to build new greenhouses or animal barns in addition to getting regular production tasks accomplished.

9. UNEVEN PRODUCTION AND SALES (*MURA*)

Mura translated from Japanese means "unevenness," "irregularity," or "lack of uniformity." In a production environment it refers to sales and production spikes and dips. Standardized and predictable work is easy to perform efficiently. A worker can readily find a rhythm, which simplifies spotting waste and making improvements. But uneven work is often inefficient because it involves less rhythm, more mistakes, and higher costs.

On vegetable and fruit farms, some amount of *mura* is unavoidable, as fresh products on such farms will ripen according to their natural season. But expanding production seasons and spreading out sales of food products—whether from animal or produce farms—has a leveling effect on farmwork and increases efficiency.

10. UNUSED TALENT

Many farms need lots of help during harvest or extra hands at butchering time and can get by with less labor the rest of the year. It's tempting to divide a farm workforce into two camps—one for workers who grind away, heads down, completing simple, mindless tasks and another for workers who think, process data, design systems, and complete complex and more interesting assignments. But to do so disrespects workers, and the farm loses out on talent as well.

Lean places emphasis on the shop floor (or *gemba*) as the best place for new ideas to generate. Responsibilities are pushed down the organizational ladder so that problems are looked at from many angles (see Chapters 8 and 9). Farm laborers working with production details day in and day out will often have better insights than the farmer on more efficient ways to get a job done. But systems need to be set up to receive and incorporate their ideas. According to lean thinking, any good idea that goes unspoken is a form of waste.

▪ ▪ ▪

While these ten types of waste are distinct categories, there is a lot of interaction among them, particularly among *muda, muri,* and *mura.* If you rush around to fill orders at the last minute, your farm is exhibiting symptoms of *mura*—uneven production. When your production is uneven, you and your workers are bound to experience some degree of overburden (*muri*), for example, when equipment fails from overuse. Mistakes are made during these times, causing defects, one of Ohno's original seven *muda*.

Once you understand these ten forms of waste, you can look out for them on your farm and root them out. This is the basic practice of lean—banishing waste to increase efficiency. However, if you want some guidance or if you want to take your leaning to the next level, read on: in the next two chapters I discuss lean tools to help with this task.

CHAPTER 5

Flow I: Tools to Root Out Farm Production Waste

The world rests on principles.

—Henry David Thoreau *to*
H.G.O. Blake, December 19, 1854

When you spot and remove wastes, you create flow—a condition in which your product gains in value smoothly and predictably, without interruptions.

Leaning on our farm started with a burst of 5S action, then a lot of research and talk—about seeing value and seeing waste. It was then time to dig in and start using lean ideas to make *transforming actions* on our farm. It was time to remove waste and to allow value to *flow*.

But where to begin? We had employed 5S to clean up the farm, but what we really wanted to do was remodel it.

We needed tools.

On the surface, waste elimination sounds simple. In practice, it can be complex. It requires a combination of careful observation, counterintuitive thinking, and hard work. Only in the second year and beyond—after we had hauled tons of junk off the property—did we really root out deeper levels of waste: production wastes. The goal of 5S, after all, is to clear away clutter in order to see the farm's flow. For example, in the case of tomatoes, after we had decluttered, we designed a germinating chamber to speed germination and reduce defect, developed yield-tracking procedures to minimize over-production, and analyzed our harvest procedures to get rid of wasted movements (see Chapter 10).

Because all farms are different and because the types of waste will vary from farm to farm, there is no magic bullet—no perfect set of three or four tools—to lean up all farms. Yet the following principles could help most farms root out a lot of waste.

Tool 1. Minimize moves.

Tool 2. Lighten the load (make every ounce count).

Tool 3. Don't overdo.

Tool 4. Ask why five times (get at root causes).

Tool 5. Employ mistake-proofing (*poka-yoke*).

Tool 6. Shorten cycle time.

Tool 7. Choose technology with a human touch.

Tool 8. Order supplies just in time.

Tool 9. Benefit from the expertise of others.

Tool 1: Minimize Moves

Your customers pay for your products and services, not your movements. For the most part, they don't care how much effort goes into your task.

Waste in a Can of Soda

In *Lean Thinking*, Womack and Jones argue that in most businesses 50 percent of time spent in production can be cut in half almost immediately by applying lean flow principles. They write, "It's hard for most managers to see the flow of value, and, therefore, to grasp the value of flow." Once managers do start to see value versus waste, they are ready for some hard work to root out the wastes, since the effort is worth it. "Flow principles can be applied to any activity and the results are always dramatic," Womack and Jones assert.[1]

Especially in mass production, wastes can be staggering. In one study of a cola maker, the team created a value stream map of one carton of cola. They found that the amount of time actually spent creating value was three hours, minuscule in relation to the total production time of 319 days! The reason? *Muda:*

More than 99 percent of the time the value stream is not moving at all: the muda of waiting. Second, the can and the aluminum going into it are picked up and put down thirty times. From the customer's standpoint none of this adds any value: the muda of transport. Similarly, the aluminum and cans are moved through fourteen storage lots and warehouses, many of them vast, and the cans are palletized and unpalletized four times: the muda of inventories and excess processing. Finally, fully 24 percent of the energy-intensive, expensive aluminum coming out of the smelter never makes it to the customers: the muda of defects (causing scrap).[2]

In many ways the work of farming boils down to moving objects from one place to another. Any time you can eliminate a move, you get rid of motion waste. One afternoon Brenneman watched us pull in a harvest, wash our food, and distribute it to our customers. He pointed out a lot of moves that we could have avoided. We went into the walk-in cooler too often with small loads. We stumbled over each other as we brought food in and washed and packaged it. We closed and opened doors too many times. Many of our moves were habits. We'd never put them under the microscope of lean.

USE SPAGHETTI DIAGRAMS TO SHOW LINES OF WORK

To get a sense of our moves, Brenneman imagined watching our farm from overhead and traced a line on paper whenever people moved around. The result: our farm on paper looked like a plate of spaghetti. I suggest trying it on your farm. Ask someone to watch you work for an hour and trace lines of movement. The drawing might surprise you.

On farms there are three ways to minimize moves.

1. The first way is to *straighten* noodles. Whenever you can move an item in a straight line rather than around a corner, you shorten steps and

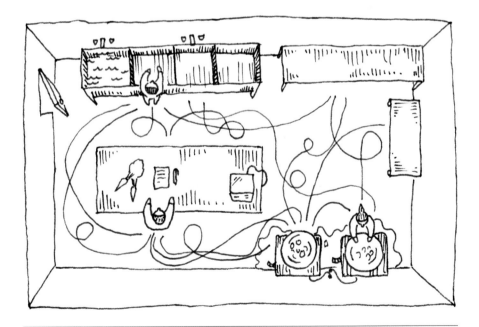

Spaghetti Diagram

remove motion waste. Think about driving a car. Anytime you turn you lose momentum and increase your time on the road.

One of the best times to create straight lines is when you are laying out your property. Keep buildings and workspaces as close as possible, and line pathways so that produce, fruit, or animals have direct routes from one production area to the next. Give transport vehicles easy access down short lanes to storage buildings, coolers, freezers, or grain bins. When we laid out our farm, we lined up all four greenhouses and built a straight lane connecting them to the processing area just a few steps away.

We recently redesigned the processing area to address the problems Brenneman saw. Before, we washed in an open yard and our setup changed week to week. We circled around each other a lot. Now produce enters our wash area at one end, passes through one or two wash stations (the second is heated for working in the winter), then proceeds smoothly to a walk-in cooler. Our truck can back up to the cooler for quick loading.

2. A second way to minimize moves is to *shorten* noodles. To keep noodles short, we decided to decrease the size of our growing area. We used to grow on plots that were a hefty walk from the main part of the property. Now we pack our food into tightly seeded beds so we don't have to walk as far when we tend our crop. No crop is more than 500 feet from the processing room.

Steve Lecklider at Lehman's Orchard in Niles, Michigan, applies the same principle on his much larger 80-acre fruit farm. When I visited him in 2014, I commented that his fruit trees and berry bushes looked like they were packed in much closer than I'd seen at other orchards. He explained that he prefers small lanes, tight spacing, and regular trellising: "We grow fruit, not trees." Lecklider went on to explain that his trellising system consisted of tying branches along horizontal lines. As a result, parts of his orchard look more like a long hedge than a grove of trees. The system keeps fruit at eye level and clustered for quick harvest. Otherwise "trees sprawl out, and we have to work more to harvest," he said.

Lecklider also keeps work areas close together. His barns, processing area, and retail shop are all spaced so movements between the areas are economized. He said he is constantly aware of "lost motion," the most insidious example of which is looking for tools. He keeps them hung at eye level in the middle of the property, and everything has a home.

Apple harvest at Lehman's Orchard in Niles, Michigan. Owner Steve Lecklider uses tight spacing and a trellising system to minimize moves at harvest time.

3. The third way to minimize moves is to *eliminate* noodles altogether. You can get rid of some small noodles quickly, for instance, during hand-harvesting. To reduce moves there we asked workers harvesting baby greens to completely fill a hand before placing the greens in a container rather than harvesting the crop a few leaves at a time. Likewise with rhubarb, we asked workers to pick a full hand of five or six stalks and then trim the tops rather than picking and trimming one by one. Another example: in *The New Organic Grower*, Eliot Coleman says that in the case of tomatoes, "forty percent of the picker's time is spent moving the hands to the basket." As such, "The picking rate can almost be doubled by learning the finger dexterity needed to pick two fruits at once."[3] He explains how workers can shift tomatoes into the back palm of their hands and pick a second tomato before returning to the basket, thus eliminating a move. Cutting out even small moves shaves off hours of work over the course of a season.

Longer noodles can also be eliminated. On our farm, we analyzed our driving patterns and found ways to congregate deliveries, doing away with much of the driving. We bought a John Deere Gator (see Tool 2) that allowed us to haul twice as much food from field to barn as our old carts. Many if not most farm tasks contain unnecessary moves and can be completed more efficiently with the selective use of technology or by economizing motion. Time spent analyzing tasks for motion waste is time well spent.

HARVEST AS MARKET-READY AS POSSIBLE (USE SINGLE-PIECE FLOW)

Single-piece flow means adding as much value as possible to the item in your hand before setting it down. It is an excellent way to minimize moves. It's a simple concept but not always one's first instinct.

Moving versus Working

In the Japanese language, the character for "moving" and the character for "working" are almost identical. The difference is that working includes a mark called a radical that denotes "person." In other words, working is moving with human wisdom. Moving is motion without thought.

The words are pronounced the same, as "doe." For Ohno, this was a problem. He writes that in the region where Toyota built its factories, speakers use the words completely interchangeably, that is, "A worker moves well" and "A worker is hardworking" have an identical meaning. Ohno explains,

> We built a factory right in the middle of these people, so the employees of Toyota think that moving and working are the same thing. Because they thought that moving with a lot of energy meant they were working, I had a terrible struggle persuading these people otherwise. We should not interpret human motion to mean the same thing as human work. We need to think of motion that includes human wisdom as being something completely different from animal-like motion.[4]

Many farmers and farmworkers make the same mistake. They confuse moving around a lot on their farm or moving quickly with work. Ohno's advice: "Learn

"TO MOVE"

"TO WORK"

Womack and Jones use an illustration from the world of office management. Imagine you have 100 letters, 100 envelopes, 100 seals, and 100 stamps. Is it faster to fold, stuff, seal, and stamp them all in separate steps, in large batches, or to complete the entire process for one letter at a time?[6] As it turns out, the latter approach—single-piece flow—is in fact much faster than the former method—batch and queue—because you are saving moves. You avoid picking up and putting down each letter four times.

I remember working for a few months at a rural agriculture development center in Guatemala with a staff of around forty to fifty people. The routine was for all staff to eat lunch together, usually rice and beans. We ate on benches or sitting on the ground under the canopy of trees. When lunch was finished, people washed their own plates and silverware out of a large circular tank in the middle of the property and set their cleaned dishes on a rack. Since the tank was large enough to accommodate several washers at a to see the difference between motion without the human element and actual work. Some call this being able to see waste."[5]

In our rural neighborhood, which consists mostly of Amish and Mennonites, large tasks like siding a building or putting up silage are sometimes shared among neighbors. These "frolics" are usually accompanied by food, and they are a good time, a chance to catch up as well as to get a job done. We look forward to invitations. One of the reasons I enjoy them is that they always feature efficient work, that is, moving with human wisdom. In Plain communities, many parents teach children elements of this type of work, such as how to predict where one's skills are best used, how to bring the right tools to a job and keep them maintained, and how to think with other workers rather than just showing up and following orders. I've worked with Amish teenagers and even preteens who constantly analyze flow and look for ways to

remove impediments to smooth work. Their minds are constantly engaged.

A few years ago we needed to replace the roof on our chicken house. It was our turn to host a small frolic. We invited a few Amish neighbors, who arrived with their own nail pouches and tools. As soon as they showed up, our neighbors started thinking about the best way to move panels from ground to roof to minimize moves. We decided one person would hoist panels to the roof edge while two people on the roof would lift them up and place them where they belonged. Everyone communicated openly. As work progressed, there was plenty of banter and joking around but also astute awareness. People who were ahead in their tasks didn't sit idle and wait for orders. They looked around and pitched in somewhere else. If anyone struggled to keep pace, someone would join them to prevent a bottleneck. The work was both mental and physical. In a few hours, we had a new roof.

time, there was no waiting for the person in front of you. Back home, dishes from large meals were usually gathered into large piles and then one person picked up each dish again to wash—batch-and-queue style. The single-piece-flow Guatemalan system saved several hours each week for a cash-strapped organization because it eliminated hundreds of tiny moves.

Our minds are instinctually drawn to batch-and-queue thinking. Factories, organizations, and offices are often separated into departments that perform tasks in large batches. Managers think they are saving time, but often they are only adding moves. Batch-and-queue thinking also dominates many farms. At harvest time workers think their task is to harvest the largest batch of carrots or potatoes or apples possible, then pass the pile along to the sorters. The sorters pick out seconds and set the batch in a queue for cleaners. The cleaners assume their task is to wash and turn the batch over to the packers. Once packers have put the food into bags or boxes, they hand the pile over to the labelers.

The *motion waste* with this approach is enormous, given that a single item might be picked, handled, and set down four or five or more times. Batch-and-queue systems always add waiting waste, since food sits and waits for the next operation. Even farmers harvesting on their own use batch-and-queue thinking by harvesting piles, then moving the piles over and over again through sorting, washing, and packing procedures. While many harvest tasks need to be completed in batches because of equipment (field corn and soybeans, for example), many times this method can be avoided.

As Womack and Jones explain, "We all need to fight departmentalized, batch thinking because tasks can almost always be accomplished more efficiently and accurately when the product is worked on continuously from raw material to finished good."[7] The challenge is to see your work from the point of view of the object you are producing rather than from the perspective of your equipment, buildings, workers, organization, or anything else. What does the *object* need to move forward in a continuous, smooth manner? Surprisingly often, large batches and queues add bumps and interruptions, while single-piece flow keeps work moving ahead efficiently.

Once we'd wrapped our minds around this principle, we found dozens of ways we could use it on our farm. Most significant, we started instructing workers to harvest as market-ready as possible. In the beginning, our practice for all of our crops was to harvest, bring everything up to the processing area, and sort out the dirty mess later. Now we do as much processing in the field as we can, while the item we picked is in our hand. This means bringing containers or bags, rubber bands, twist ties, and washing supplies—everything we need to get the crop from soil to delivery box or bag—right into the field.

For example, just yesterday I received an order from a wholesale customer for twenty heads of lettuce. So I took with me to the field the customer's waxed produce box and individual lettuce bags. One by one I picked the heads, dunked them in a bucket of clean water, shook them off, placed individual bags around them (at the customer's request), and put them straight into their final box. I never set a single head down until it was customer-ready. We do the same with kale and other bunched items. We pick, sort, band, and box, all in one move. With herb packs for our CSA boxes, where we don't need an exact weight, we take small bags or clamshell packages into the field and pick directly into them.

We can't use seamless single-piece flow in every situation. For instance, when it's hot, head lettuces need to get into a cool room as soon as possible.

There is no time to process them in the field. Other items, like baby greens and microgreens, which require special rinsing and handling equipment, need to be washed in a climate-controlled processing area.

Every year we search for ways to improve single-piece flow. I am currently working on a design for a small, portable roots washer and packing station that workers can take with them directly into a bed of carrots or beets or potatoes so that small batches can be picked, sorted, washed, packed, and labeled from anywhere near a hose. To take the idea a step further, we've considered an enclosed refrigerated wagon or compact vehicle with separate compartments for our different value streams (wholesale, farmers' market, CSA), so that after they pick and pack items, workers can simply turn around and place them in the cooler wagon. We are also working on a redesign in our salad greens washing area that will allow individual workers to bag, weigh, seal, and tote all in one sequence, without setting down a bag. For years, we completed each of these steps in separate batches in assembly-line fashion with two, three, or four workers. The challenge with each of these workflow redesigns is to put ourselves into the shoes of our products, so to speak, in order to discover the smoothest way forward from field to customer.

A lean principle we follow: add as much value as possible to the item in your hand before setting it down. Here I sort and bunch as I harvest, not later.

Lean as a System: Applying Tools Systematically

Just because you have a hammer and know how to swing it doesn't mean you can build a house. In fact, even if you can grade, pour foundations, frame, run electricity, plumb, insulate, and apply siding, roofs, and drywall doesn't mean you can build a house. You need to see the big picture and be able to *integrate* what you know.

Likewise, a quick online search will reveal countless lean tools for increasing profits and gutting waste. But the use of any one tool alone is less effective than a combination of tools applied in a systematic way—always with an aim of adding more value, not just randomly cutting costs. For example, the tool "harvest as market-ready as possible" works best if you are harvesting precisely what your customers want and if you follow up by using metrics (see Chapter 6) to analyze whether the tool worked. Speeding up any one process will do little good if it's not part of a bigger plan.

Tool 2: Lighten the Load (Make Every Ounce Count)

Toyota uses an evaluative system called Toyota's Verification of Assembly Line (TVAL) to rate jobs based on the physical burden of work positions, including how much weight workers have to carry. The system gives Toyota managers objective data they can use to redesign work. Jobs with the highest ratings are examined for ways to reduce human lifting.

Lean does not mean a faster, more hectic work pace; it means smoother, more efficient motions so workers can accomplish more with less fatigue. Even small amounts of unnecessary lifting can add up to a great deal of *muri*, overburdening waste.

Eliot Coleman calculates that "an average field worker using a cultivating hoe will make some 2,000 strokes in the course of an hour's work."[8] Tools should be as lightweight as possible. Even a few more ounces than necessary, he says, means a lot more lifting. Coleman is right, and he is thinking lean. If a hoe can weigh half a pound less (Coleman says no hoe should weigh more than 1½ pounds), and if you raise that hoe 2,000 times in an hour, that equals 1,000 pounds that you didn't have to lift that hour.

At first we didn't think much about the awkwardness of our tasks. We just wanted to get them done. Now we scour our farm for *muri* all the time. We don't used a numeric TVAL system like Toyota, but we do take note of

Lean says to analyze your work for burden. Lightweight, well-designed tools, such as this long-handled hoe, remove overburdening waste. Photo courtesy of David Johnson Photography.

when people and equipment are overburdened and quickly work to remove the *muri* by redesigning those jobs or replacing inefficient equipment.

For example, the hoes we started with were an eclectic mix of heavy, poorly designed tools we'd picked up at auctions. We got rid of nearly all of them and invested in smaller, lighter tools. I switched from a heavy multi-purpose tool to a carbon-bladed Opinel pocketknife with a lightweight wooden handle. I've even updated my wardrobe from heavy canvas, carpenter-style pants to lightweight hiking pants with articulated knees, so there is no friction when I kneel down to harvest.

Another example: When we started our CSA ten years ago, we used to ready our deliveries by laying out empty boxes and then walking around them with heavy produce totes, dropping items into each stationary box. We eventually switched to cafeteria-style loading: heavy totes stayed on a table while workers walked through the line with CSA boxes, pulling one item from each tote. This allowed us to keep our heavyweight items (the full produce totes) stationary and instead move the lightweight items (the CSA boxes). One day an intern suggested lowering by 4 inches the front part of the table, where the CSA boxes slid. Now workers use gravity to lower items into the boxes with less reaching.

These little steps toward efficiency seem small, but together they have trimmed thousands of pounds of weight and lots of *muri* from our workload.

We've found several ways to lighten our load at harvest time. One is by using a custom-made root digger to loosen up crops like carrots and garlic. Formerly, we had to wrestle each carrot and head of garlic out of the ground by hand. I explained our problem to our local Amish machinist, and together we designed an adjustable-height and adjustable-tilt root digger that fits perfectly behind our small tractor. The blade tapers toward the ground (it's a repurposed steel piece from the end of a skid loader bucket), helping it to dig in easily. It gently lifts crops so workers need only to go along and pick them up. It has easily cut our harvest time on root crops in half and has made the work easier to do.

The machine that has removed the most *muri* waste from our farm, however,

We lowered the front part of this table by a few inches to make it easier to fill CSA boxes. Small steps like this remove a lot of burden over time.

is the tilt-bed John Deere Gator that replaces the garden carts we relied on in the beginning. We use it to haul just about everything, including tools, compost, and harvest crates. The dump feature makes unloading a snap. It is big enough to handle a hefty load but small enough to squeeze into the greenhouse. We equipped it with a shade cloth so we can also use it as a summer harvest vehicle, and we keep a basic tool kit in the glove compartment.

Muri is like a doorway that wasn't installed quite right: sooner or later, either the threshold or the door will wear down. On farms, both tools and people can wear out fast if work is not designed with smooth flow in mind.

As part of our leaning, we widened our greenhouse openings to accommodate equipment (left) and employed a custom-made root digger (right) and tilt-bed John Deere Gator (bottom) to remove thousands of pounds of lifting from our work. Gator photo by Emma Gerigscott/Clay Bottom Farm.

Tool 3: Don't Overdo

Your goal in lean production is to give your customers exactly what they value, not *more* than they value. Overpleasing is a type of waste.

DON'T OVERPACKAGE

When we first started selling food to customers, we knew we had grocery stores and their slick packaging to compete with. However, whenever we've tried mimicking how grocery stores adorn their goods with brightly colored plastic bling, we usually end up regretting it.

We once bought bags cut in the shape of a head of romaine lettuce, with air-breathing slits on the sides and the word *Romaine* printed in bold letters down the middle. But we sold our romaine faster in cheaper plain bags. Our customers didn't care. Concerned about the need to brand, we once bought a laser printer and printed out fancy colored labels, complete with our logo, to slap on many of our items. But when we removed the labels, our food sold just as fast.

I've heard the same story from other farmers. Randy Ewert at Bair Lane Farm in Marcellus, Michigan, told me that for all of his mixed greens bags he used to print labels that included a complete ingredients list. He scrapped the customized list and now uses a more generic label. "Our ingredients change all the time with the weather, and our customers didn't really care," he told me.

This isn't to say there is no place for branding, labels, and packaging. Our sales of greens went up once we started selling them in smaller ¼-pound bags in addition to ½-pound bags, and we know farm visitors value the free "Eat Local Food" stickers we hand out. But you should package only as much as is absolutely necessary to sell your product.

DON'T CLEAN TOO MUCH

We use the customer-first barometer: What is important to our customers to keep clean? Clean that much, then put the broom down and do something that adds value.

Many customers tell us they choose our food because it's prewashed. We hose and wash food to their standards and sanitize surfaces that come into contact with food: totes, wash basins, and prep surfaces. I never clean the lawn mower or skid loader or my implements more than necessary—just enough to keep them functioning well. We oil our hand tools, but we don't polish them. Dust and dirt are always on the hood and in the crevices of my tractor, but as far as I know my customers don't care.

For cleaning that we must do, we find ways to do it efficiently. We installed high-pressure lines in our processing area and store cleaning solutions at eye level. We hang brooms and dust pans close to pads that we sweep regularly. Totes and crates that need regular washing are permanently stored next to the wash station. When we come home from a market with dirty containers, we can back right up to the wash area for one-step unloading and washing.

Big lawns are a baffling form of waste on a lot of Midwest farms. How is a 5-acre lawn adding any value to your product? If you are fascinated by open landscapes and use your riding lawnmower for Zen meditation, then mow away. But in most cases you are serving yourself, not your customer. To lean up our mowing, we fenced off large areas of grass for use as pasture. We keep a smaller yard mowed for neighborly appearances but make a point to grumble every time we mow it. Maybe someday we'll finally plant perennial beds or get sheep to do the grazing.

Tool 4: Ask Why Five Times
(Get at Root Causes)

Ohno instructed Toyota engineers to ask why five times whenever a problem occurred: "By repeating *why* five times, the nature of the problem as well as its solution becomes clear."[9] The point is to get to the root of a problem so you can fix it at the source. Interestingly, almost all little children do this by instinct—they ask why repeatedly, out of an insatiable desire to understand root causes—yet adults generally find the practice tiresome and annoying.

In production environments, we do well to mimic our children. In my experience, whether you ask why five times or two times or twenty times isn't important. What matters is that you ask the question until you get to the bottom of your problem.

A couple of examples:

Problem: A customer finds a rotten pear in her bag when she gets home from the farmers' market.

1. Why? Because booth staff did not notice the rotten pear.
2. Why? Because the pear was at the bottom of a bag.
3. Why? Because sort room staff placed it there.
4. Why? Because staff were not trained on proper grading standards.

5. Why? Because the farmer forgot to mention the standard during a training.

> The root problem of the rotten pear goes back to training protocols that were not written down, and that's where the problem should be solved.

The following is an example from our own farm several years ago.

Problem: Weeding in the winter greenhouse is taking too long.

1. Why? Because there are a lot of tomato weeds (tomato seeds that germinated and grew).
2. Why? Because tomatoes rotted on the ground during the summer.
3. Why? Because workers didn't pick up rotten tomatoes.
4. Why? Because it would have taken too long during a busy time of the year.

> So after four whys we decided to start using landscaping fabric under our tomatoes. Now seeds never find their way to the soil, which speeds up weeding in the winter.

Here is another example from our farm:

Problem: Mesclun mix yields out of the greenhouse are low.

1. Why? Because seeds germinated poorly.
2. Why? Because overhead sprinklers underwatered.
3. Why? Because the timer quit working.
4. Why? Because its battery ran out.
5. Why? Because replacing timer batteries never made it to the yearly calendar.

> In this case, the root cause of low yield is a problem of maintenance scheduling.

In many ways lean is about applying the scientific method to your production problems. The scientific method relies on verifiable evidence (facts) to test—and modify, if need be—hypotheses about the way the world works. The work of a scientist is a continuous cycle of theory (hypothesis) development, testing and analyzing results, and formulating new hypotheses. Asking why is the first step scientists use to develop a hypothesis. Likewise, lean farm managers should ask why two times, five times, or twenty times to get at root causes and constantly generate ideas about better ways to do things. Then they should rigorously test their ideas with real data and finally modify their products or approaches.

In short, gather the facts, go to the source of the problem, tweak your process, then test again.

Tool 5: Employ Mistake-Proofing (*Poka-Yoke*)

Another way to rid your farm of defect waste is to design quality-check routines into your work flow. The best way to create a *poka-yoke*, or mistake-proof, condition is to check yourself regularly in order to find defects when they first occur, not when your customer discovers them.

At Toyota, engineers design *poka-yoke* into automated machines. For instance, a laser can sense if a tool bin is empty and stop production before machines keep working and cause errors. Toyota factories were once known for the andon cords placed above workstations. When workers saw a defect, they would immediately pull the cord to stop production. Although those cords are now being replaced by buttons—a reduction in overhead clutter—the system remains. Some factories use a three-light setup. Ohno explains, "When operations are normal, the *green light* is on. When a worker wants to adjust something on the line and calls for help, he turns on a *yellow light*. If a line stop is needed to rectify a problem, the *red light* is turned on." Defect prevention is everyone's job: "Workers should not be afraid to stop the line."[10]

When we first started out, we were too afraid to stop the line. We were often in such a hurry to get our product out the door that we cut quality-check corners. As part of our lean up, we implemented systems to spot defects before our customers did. Primarily, we start each day with a walk to monitor for pests or other causes of defect so we can catch them early. We established a system to double-count CSA boxes so that we are sure the correct number of boxes is going to each location. Ideally, boxes never leave until someone other than the person who loaded them double-counts. The delivery person counts boxes coming off the truck and writes the number on a sheet that we keep at pickup locations.

At Bair Lane Farm, Ewert noticed that no matter how careful they were to count their CSA boxes, one often seemed to go missing at their pickup location. Now he always sends along one extra CSA box to preempt a mistake. If mistakes happen repeatedly, find systems that automatically catch or prevent them.

Tool 6: Shorten Cycle Time

In a lean enterprise, the goal is to produce your product in as little time as possible. This increases worker productivity per hour and ensures that you are getting the best use out of your infrastructure.

But on farms can we really shorten cycle time? Don't tomatoes always take about 100 days to grow? Don't grassfed steers always need two years to reach butcher weight? In recent years science has gained a lot of ground in shortening agricultural cycle times. Some of these techniques go too far—growth hormones in cows, for instance—and represent examples of lean overreaches (see Chapter 12). However, there are less invasive ways farmers can shorten cycle times and also keep farms in a healthy alignment with nature. Here are three such options:

1. *Choose shorter-season plant varieties.* Non-GMO options do exist. In vegetable farming, seed breeders are constantly developing shorter-season varieties. The shorter time a crop is in the ground, the less chance for defect and the faster you can get it to market. In the high-value space of our greenhouses, we purposely choose fast-maturing varieties. For example, we don't grow poky crops like cauliflower, storage carrots, leeks, or brussels sprouts under greenhouses because they eat up greenhouse space (a fixed cost) for too long. Instead, we focus on radishes, short-season carrots and turnips, mesclun mix, spinach, Asian greens, pea shoots, and microgreens, which sometimes take less than a week to mature.

2. *Discover processes that speed growth.* We noticed one winter that minimal supplemental heat in greenhouses greatly speeds growth. So now we keep two winter greenhouses heated to only 30 degrees F (-1 degree C), which doubles plant growth compared to unheated greenhouses. Every year we also use more transplants—rather than direct seeding—to speed production. Many crops, like spinach, turnips, and beets, can be successfully direct-seeded, but unless soil temperatures are perfect, germination time can be painfully slow. We've waited months for some seeds to germinate. With our farm-built germination chamber and transplant system, we keep crop growth rates optimal, sometimes shaving weeks off a crop's life cycle.

   Another example is from Blue Heron Farm, in Millersburg, Indiana. Adam Derstine, who runs the farm with his wife, Elise, and friends Tom Stinson and Cassandra Byler, told me he notices that their pigs grow faster when they have the right kinds of forage available. A recent litter didn't have ideal pastures, and the pigs slowed down. They continue to experiment with different forages like brassicas, canola, and sorghum to see which ones help pigs mature more quickly.

3. *Cut out wasted steps after plants or animals have reached maturity.* Post-harvest bottlenecks lengthen cycle time. One bottleneck on our farm used to be filling tanks for washing greens. We use a stainless steel four-bay sink. Each bay is about 24″ × 12″. The faucets that came with the sink fill the bays in about ten minutes. We fill the bays twice a week, so that's twenty minutes of waiting waste every week. I replaced the faucets with high-pressure hoses that fill the bays in about two minutes. This reduced waiting waste by sixteen minutes per week. Likewise, the bays used to drain slowly. After I replaced and widened the drain, they now empty out in just a few seconds. These changes shortened the timeline for our greens. With rare exceptions, your goal should be to get your product into your customer's hands as soon as it is ready. The longer a product sits on your farm, the longer its cycle time and the higher your costs.

Tool 7: Use Technology with a Human Touch

Lean is not antitechnology. But because technology applied blindly, with too much exuberance or for the wrong goals, will lead to work waste in addition to unnecessary environmental burdens, lean places an emphasis on choosing equipment that is the right size—human-scaled—and combining human intelligence with the work of machines. The right tool can remove enormous burden from workers, but too often complicated machines that were not carefully engineered or scaled for their task only add burden.

Machines should be selected with careful thought. They should fit into the farm and serve the needs of humans on the farm, not the other way around. Lean thinking compels farmers to ask pointed questions before purchasing new equipment. Will the new machine further the farm's cause and reduce farm waste, or will it compel farmers to change what they do in order to conform to the machine or to pay it off? What new wastes might a new technology introduce? Is the new machine simple enough for anyone to use, or will it require burdensome training? Will the time required to service and maintain the machine pay off? What about customers? Does the machine allow us to increase value for customers, or does it simply add to our costs and theirs?

As part of our leaning, we analyzed our equipment and made several changes. Below are ways we have tried to keep a human touch with our equipment.

CHOOSE THE RIGHT SIZE

There are several reasons to get size right. First, equipment is usually a fixed cost, and in lean the goal is to maximize your fixed costs (see Chapter 6). If a tool or machine is the right size—big enough to do the job well but no bigger—it is more likely that you will use it to its fullest capacity and get the most for your money. Tractors are a great example. Why own a $40,000 tractor when a $15,000 tractor would suffice? You're just locking up $25,000 in unused infrastructure capacity. One decision we have never regretted is the purchase of a small tractor. We traded in a Ford 8N for a 34-horsepower Kubota compact tractor. We can now do prep work with less effort. But the tractor is not overkill: it does just what we need it to do, which allows us to make the most out of our investment.

Another benefit of small and simple equipment is that more people on the farm will be likely to use it. We have trained many workers to use our tractor and implements. For those who have never worked with machines, we sold a larger, awkward walk-behind cultivator and replaced it with a small Troy-Bilt Jr. tiller that just about anyone with a few minutes of demonstration can operate. It is powerful enough to accomplish tough jobs but simple enough for new workers to use. Technology that serves human needs well will be engineered for ease of use as well as function.

Small equipment is also easier to use in tight areas. Our skid loader is large enough to handle big loads but small enough to maneuver anywhere we need. We replaced a tractor-mounted sickle mower with a smaller one that mounts to our BCS tractor because it can handle both large and small areas. Our tractor can fit

Human-scaled but powerful equipment, like this Troy-Bilt Jr. tiller, can perform jobs well but are simple enough for anyone to use. Photo by Emma Gerigscott/Clay Bottom Farm.

between greenhouses, and with its skinny profile and tight turning radius, it allows us to keep pathways narrow.

By keeping equipment small, we also save on maintenance and repair. In general, the more expensive a machine is, the more costly it will be to maintain and repair. There is no point in maintaining a complicated or bigger machine than you need.

In short, if your market and scale don't call for mass-production equipment, then don't use it. The principle sounds simple, but machines that are too large and expensive are a common source of waste on farms. Better process, not bigger machines, is what many farms need.

STAY FLEXIBLE: USE TOOLS AND MACHINES THAT PERFORM MANY JOBS

Whenever possible, we prefer equipment that can perform multiple tasks, so that one tool can take the place of several. In Chapter 1 I talked about how we lined up all of our hand tools, chose a few that we decided could accomplish what we needed, and got rid of the rest.

The same thinking can apply on larger farms. Too often large-scale corn and soybean farmers find themselves without options when commodity prices drop because they have invested so heavily in expensive, single-use equipment—their equipment restricts what they can do. These farmers would do well to carefully gauge their investment in such equipment and whenever possible seek out equipment that will perform many jobs. Agricultural engineers should help them out by designing products with more flexibility in mind.

When I asked Brenneman how he makes machine-purchasing decisions at his trailer company, he told me that he is wary of the highest-end technology. In his experience, expensive, dedicated machines rarely serve his needs. He told me he would rather have a $100,000 tool that is highly flexible, though perhaps less efficient or automated, than a $200,000 tool that can run on its own but is a one-trick pony, performing only a single task.

AUTOMATE BUT KEEP TABS

Automating systems can save hours of labor. But even with the best technology, malfunction happens and defect waste can result. That's why lean managers keep close tabs on technology. They mistake-proof their machines. At Toyota, machines tend to be engineered to shut down automatically before they produce defective products in mass quantity, as with the laser beams to detect when parts bins are empty; an empty bin triggers machines to shut off. One concrete example from our farm: while we rely on automated heating and ventilation for our greenhouses, we also use a weather monitoring system to keep track of greenhouse temperature in case of equipment failure. The system sends us a text message when temperatures are outside of the range we want. Technology offers benefits to farms of all sizes, but it should be used cautiously, with strategic intention, and always with a human touch in mind.

Tool 8: Order Supplies Just in Time

Too much inventory of raw materials on hand is costly. The accounting world calculates these costs as inventory carrying costs. They include, among others:

- Costs of warehousing (taxes and insurance on buildings, the cost of stumbling around objects)
- Costs of potential damage
- Costs of depreciation
- Costs of obsolescence (a part lingers so long that it's no longer useful)

In addition to all these costs, inventory binds up capital that could be used to create value. Why buy more supplies than you need when you could use your money for more pressing investments?

Toyota used the idea of "just in time" to keep inventories as low as possible. Rather than stockpile parts, they demanded that suppliers deliver just the right amount just when the item was needed (within hours of actual assembly), avoiding tremendous inventory carrying costs. In addition to saving space and money, the practice allowed for small-batch production, which meant Toyota plants could change what they were doing very quickly. Many manufacturers are burdened by piles of obsolete parts, and they sometimes design goods to use up that inventory. They let last year's poor decision to overbuy parts guide current production. Not so for Toyota. Their production was timely and precise. It was guided by the exact item customers wanted, not by parts supply levels. The practice gave Toyota incredible freedom, though it required careful communication with suppliers. According to Ohno:

> There is nothing strict about what we do. Whether the delivery is from a supplier or from another Toyota department, if we need it at one o'clock we require delivery by eleven o'clock. All we are saying is that delivery by nine o'clock is too early.[11]

Womack and Jones describe a visit to one Toyota factory in which "there was no parts warehouse at all." Instead,

> *Parts were delivered directly to the line at hourly intervals from the supplier plants where they had just been made. (Indeed, our initial plant survey form asked how many days of inventory were in the plant. A Toyota manager politely asked whether there was an error in translation. Surely we meant minutes of inventory.)[12]*

While farms are not assembling parts together as factories do, there are many examples of inventory waste on farms. To name a few:

- Seeds sitting around and losing viability
- Too much animal feed on hand
- More supplies—row covers, hoses, and plant stakes—than needed
- Too many bags and labels
- Too many packaging boxes

The temptation for farmers when ordering supplies is to save money on shipping by putting in large orders at the beginning of the year. But it is impossible to know how many seeds or other supplies you will need as the year progresses. The cost of keeping too much inventory usually quickly erases the savings.

We've made the same mistake plenty of times. At some point we've overordered seeds, potting trays, composts, potting mixes, and just about every other supply we use. As part of our leaning, we now keep closer tabs. Two of our suppliers have taken steps to make it easier for farmers to order just in time. One offers free shipping on all orders with no minimum order, and the other prides itself on prompt (often same-day) shipment and sends a list of varieties and amounts ordered in previous years. Other suppliers also increasingly recognize that farmers save money when they order more frequently, in lower volumes. We now order our seeds as we need them rather than once a year. With rare exceptions, we never carry seeds over from one year to the next because the defect waste in seeds that don't germinate is costlier than throwing out the occasional extra seed packet.

We've taken other just-in-time steps. We now keep a minimal amount of row covers on hand and buy the shortest roll size possible when we reorder. We keep only a couple months' supply of produce bags on hand. We don't stockpile rubber bands and twist ties but order as we need them. We order compost-making ingredients (bulk organic matter like leaf mold) five or six times throughout the season rather than in one big delivery.

For some of our supplies, we use replacement cues to tell us when it's time to reorder. In lean systems these are called *kanban* signals. A simple illustration is from the era when milk was delivered in bottles. If customers wanted another bottle, they set out their empty bottle to signal the delivery person. In similar fashion, Brenneman's factory uses the two-bin system, with replacement parts kept in two identical bins. When one bin is empty, the second bin is pulled forward so workflow is uninterrupted. The empty bin—the *kanban* signal—is set on a cart and replenished by a worker assigned to refill all the bins in the factory.

An example for us is seeds. I organize seeds by keeping them in plastic bins that contain similar types—for example, salad mix or root crops. When I run out of seeds, I put the empty seed packet on top of the bin. Once a few packets accumulate, I call the seed company to order more. The practice is simple, visual, and helps us keep inventories in stock in small amounts. Another example is plastic bags. We usually buy three or four boxes at a time. We write "reorder" on the last box so that when we get to it we're reminded (or a worker reminds us) to order more. These systems help keep inventories at the lowest volume possible but never running out.

The tag room at Van Belle Nursery, Abbotsford, British Columbia. Photo courtesy of Van Belle Nursery.

An example of inventory reduction on another farm is Van Belle Nursery's tag system. In nurseries, wasted tags are a perennial problem. Gerson Cortes, a lean consultant who works with Van Belle Nursery, writes that when he visits a farm, he asks growers if he can visit their tag room. "What we see is typical—shelves of tags that are no longer used, because the variety is no longer sold. Eventually the grower will go through and dump all of the obsolete tags to make room for new tags," he explains.[13]

Tags in nurseries are usually printed out of house in large volume several months ahead of an actual customer order. Like other nurseries, Van Belle was throwing away a large number of unused tags every year, costing time and money. Their solution was to print their own. Dave Van Belle, the owner, explains, "With the exception of reprinted, branded tags, we have zero tag waste." They coordinate printing with actual need, so that "tags are printed just two hours (sometimes less) before they are put on the plants," says Van Belle. "This means maximum flexibility for us. We can change our minds on what plant to ship right up until printing."[14]

Tool 9: Benefit from the Expertise of Others

In many cases, others can more efficiently complete farm tasks than you can. My advice: let them. If you have connected a product to paying customers who value it, you've done the big-picture work of structuring a business. There is no shame in farming out pieces of your production where you lack capacity. The right partnership can eliminate many types of waste.

For example, I decided to farm out most mechanical work like tractor maintenance and complex electrical and plumbing projects to professionals, who can perform these tasks much faster than I can. When I don't know what I'm doing, there is a lot of motion waste. I end up redoing a lot of work that was beyond my abilities. Also, when jobs are done on time, I eliminate the waste of waiting. And when they're done right I avoid the waste of defect.

We also hire trucking services to deliver our compost ingredients in large loads rather than trying to move it ourselves. At times we've contracted with other growers to start plants for us. Another example: after a few failed attempts at farm-built greenhouses, we decided to use tunnels engineered by professionals. Even then, one engineered greenhouse blew away because we hadn't anchored it well enough. Fortunately, it was a smaller, experimental tunnel. The next two greenhouses we put up ourselves successfully, but the job took weeks longer than we'd projected. For our final two greenhouses, we hired a crew to help us. We knew by then how much work was involved, and how much risk, and we calculated that our time was more profitably spent *using* greenhouses than building them.

In his book *Gaining Ground*, Forrest Pritchard, a cattle, hog, and chicken farmer in Virginia, describes how he put this principle to use on his own farm. Frustrated with the high costs and hassles of making hay, he took a bold step: he gave it all up, sold all of his equipment, and bought *all* of his hay from others. The move jump-started his farm. He suddenly had time to raise animals, and he freed up capital that had been bound up in equipment. The story shows how one quick decision can eliminate a slew of waste: overburden (the hard work of haymaking), inventory (the cost of keeping hay and equipment on his property, not someone else's), waiting waste (his farm sitting idle while he made hay and fixed old equipment), and defective products (because his haymaking was not always successful). Pritchard set stubborn self-sufficiency aside and value started to flow.

Remember the carrot journey from Chapter 3? Only a few activities added real value to the carrot. Only seeding and a few harvest and post-harvest tasks tightened Shingo's bolt. Those tasks are your job. Spend your time there. The rest—the movement—is frequently better left to others.

<hr/>

Just as waste takes many forms, then, lean offers many tools to get rid of the waste. Sometimes, like Pritchard, we eliminated waste in large chunks, as if cutting it out with a hatchet: when we purchased our John Deere Gator and our tractor and when we hired help to build two greenhouses. Other forms of wastes were smaller, removable by scalpel: replacing heavy hoes with lightweight hoes, laying down landscaping fabric under tomatoes, and widening drain holes in sinks.

The point, however, isn't to ponder waste and determine its size —to calculate it, stare at it, or define its parameters. The point is to chip away at it, in chunks or slivers, until you're left with pure value.

CHAPTER 6

Flow II: Tools to Root Out Farm Management Waste

Simplicity, simplicity, simplicity! I say, let your affairs be as two or three, and not a hundred or a thousand; instead of a million count half a dozen, and keep your accounts on your thumb nail.

—HENRY DAVID THOREAU

The only way to generate a profit is to reduce cost.

—TAIICHI OHNO

Costs do not exist to be calculated; costs exist to be reduced.

—TAIICHI OHNO

In our culture of accumulation, business owners are prone to be addicted to constant expansion: bigger every year, more sales, more goods, more employees. With a farm, that philosophy can be destructive: farmers strip and overuse land, stuff animals into tighter and tighter quarters, and work unsustainable hours. Food quality suffers, the environment suffers, and people suffer.

On the other hand, a farm that is the *right* size and well kept is a beautiful sight: animals have room to roam on fresh grass and in open air; crops are healthy and well tended; a farmer's schedule is balanced and sane. Food quality is high because quality and value matter as much as quantity.

At the time we started applying lean, we did not want to add hours to our workload. We didn't have a vision to expand. We were happy with the size of our farm. But we wanted to make our work *easier*. We wanted to travel more and have more time to spend with family and friends. Too often we felt trapped by the farm. After we organized with 5S and leaned up our production using the tools from Chapter 5, we saw an incredible difference.

Leaning up increases efficiency and also creates a more pleasant farm.

We were working less, and our food quality was increasing. Our farm was indeed a more pleasant place to work.

But as we planned for the long term, we wondered if lean had management tools to offer as well. Were there ways to grow even more profitable in the next five or ten years without getting bigger? Brenneman told us about three such tools: produce only what you know will sell, cut costs to grow profit margins, and replace low-dollar-value items with high-dollar-value items. These strategies maximize fixed costs, allowing us to get the most out of our small farm and the money we've invested in it.

The target of all these tools is waste. Long-term farm management wastes are less tangible than day-to-day production wastes. Production wastes, the subject of Chapter 5, concern everyday operations. They often linger close to the surface: drain holes that are too small, crops that don't grow well, lanes that are too long. We can see them with our eyes and touch them with our hands. Long-term management wastes are a bit more elusive. These wastes occur during planning stages, when you plot out your farm's future or plan what to grow in the next year. Examples of management waste include choosing the wrong crops, producing too much or little of a product for sale, or bogging down your farm with experiments. These originate in the management office, behind the scenes, not in the field. But they are just as pernicious. They create friction, block flow, and slow down your farm. Seven tools help root them out:

Tool 1. Practice production control (stop hoarding).
Tool 2. Cut costs to grow profit margins.
Tool 3. Replace low-profit items with high-profit items.
Tool 4. Maximize fixed costs.
Tool 5. Level the load (*heijunka*).
Tool 6. Use metrics to measure your work.
Tool 7. Balance creativity and discipline: the 15 percent rule.

Tool 1: Practice Production Control (Stop Hoarding)

There will always be some guesswork involved with farming, since farmers can't perfectly predict markets or yields. There are too many variables, such as the weather, that are out of the farmer's control. But the more accurate you are, that is, the less you overproduce and the more you supply just what the market wants, the better. In the first few years of a farm, you need to try out new crops or animal products to test your markets. But after that, profitable farming requires discipline to produce the right amount. This is surprisingly hard to do.

Taiichi Ohno suspects that the reason we are addicted to such habits as overproducing is that humans are, after all, an agricultural people: we have strong instincts telling us to hoard.

Long ago, your life depended on the amount of grain you put up during harvest for leaner times. Scarcity was often the rule. It made sense to produce as much as possible and to harvest every last kernel of corn or grain of rice. For this reason, explains Ohno, "In our gut, we must enjoy inventory management more than production control."[1] We feel more secure with plenty of food around us.

Rachel and I still preserve a lot for ourselves. We put up extras for winter, freezing strawberries or canning dilly beans. We keep a stash of potatoes, squash, onions, and sweet potatoes in our home storage room. We've started growing our own corn for cornmeal. But business economics are different from home economics. The role of a business is to produce for others in exchange for cash. Business people are not rewarded for producing an overabundance and holding onto extras. They are punished. Their net profits suffer. Reining in means fighting our instincts. As Ohno writes:

> *Rather than doing proper production control upfront, we prefer to stay busy making things and then later spend effort managing inventory when "the parts were made." Books titled*

"Production Control" do not sell very well, but when you write books on "Inventory Management" they sell very well.²

Every spring, when the seeder is in my hands, I have a hard time stopping. Even though I know better, I always feel an urge to keep pushing the seeder an extra 5 feet. The problem, if I give in to instinct, is that I am wasting 5 feet of prepped ground. At weeding time there will be 5 feet more to weed. If nights are cold, there will be 5 feet more to cover. I am only adding waste.

When harvesting, I feel the same urge to hoard. I want to harvest it all, even if I know I don't have a market. I remember more than once overharvesting tubs of vegetables too nice to let go to waste. After rounds of phone calls, the food more often than not didn't sell. It turned into compost. While the food was not technically wasted (it turned into compost that fed the farm), a lot of effort was.

START CONTROLLING (FORECASTING)

Lean encourages production control to produce the exact right amount instead of as much as you can. A primary tool used to get volume right is

Opportunity Losses and the Historical Context of Lean

To understand production control ideas better, it helps to understand the context in which they originated in Japan.

Ohno explains that after World War II, in the 1950s and the 1960s, Japan suffered enormous opportunity losses because it did not have the infrastructure capacity to produce items fast enough for a rapidly growing economy. Later this led many factories to overcorrect and start using more machines and bigger equipment— to overproduce out of fear of more lost opportunities. As Ohno writes:

The word "loss" can be an actual loss or an opportunity loss, and are

[sic] very different things, but in either case people tend to feel they have suffered a great loss. I think this is another example of a misconception because the lost opportunity to make a profit causes no actual harm, while an actual loss causes financial harm. People confuse these things.³

"Don't fear opportunity losses," became one of Ohno's mandates at Toyota. Fear should never be an excuse to overproduce. The occasional unexpected sale rarely makes up for overproduction waste.

forecasting—tracking sales from year to year and adjusting the amount you produce accordingly. In lean factories, cycle times are often so short, and workflow so efficient that forecasting becomes unnecessary: units are simply built to order. On farms at least some forecasting is unavoidable because most plants and animal have long cycle times (several months or even several years) determined by nature, not human engineering.

After a few years of composting too many unsold crops, we asked ourselves what would happen if every seed germinated, grew well, and sold for a fair price? In other words, what if *every* seed turned into cash? Not half, not 80 percent, but 100 percent. That goal keeps us motivated to search for reasons seeds don't germinate or grow well (asking why five times). And it motivates us to forecast well.

Forecasting is hard work but absolutely worth the effort. The best way to forecast is to presell, as in the CSA model, which was a Japanese invention. CSA, which involves customers paying at the beginning of a growing season, eliminates almost all overproduction waste because the farmer knows exactly how much to produce. Any time you make a sale before you spend money on production, you create ideal conditions for controlling production.

Our market base includes more than the CSA, however. So we developed systems to track and predict sales. In the beginning I remember trying to keep track in my head how many pounds of each crop we sold from week to week and adjusting the next week's harvest accordingly. At seed-ordering time we mostly purchased random amounts, based on our gut feeling of how much we might need. Our forecasting was crude. Now, using the three forecasting practices below, I know with precision how many pounds of tomatoes we will sell on the Fourth of July and exactly how many tomatoes to plant, as well as when to plant them, in order to achieve the right-size harvest for that market. Ways to forecast on the farm:

1. *Precisely track week-to-week sales.* The best predictor of future behavior is past behavior. There is no substitute for tracking sales as they happen as the best indicator of how the same crop will fare in next week's or next year's marketplace. We keep track of our farmers' market sales using a Google Drive spreadsheet that a market worker fills in at the end of each market. We rely on it from week to week. We refer to the trajectory of sales over the past several weeks to project the coming week's harvest.

2. *Track production inputs needed* to produce the volume sold. Once we project how much we need, we need to know what it will take to produce that amount. In the case of carrots, for example, I need to know how many seeds, how much fertilizer, and how much land will be

If You Do Have Extras...

Not selling a crop you've produced is like a marathon runner quitting after mile 25. Production control, using tools like forecasting and precise harvesting, prevents this from happening. But if you do have extras, there are options besides tilling in, composting, or disposal. Fruit orchards perhaps have the biggest challenge with production control. Yields can be nonexistent one year and a bumper crop the next. For overproducers, the following techniques are a must.

Value-Add Leftovers

Value-adding crops, such as turning leftover strawberries into jam, can be a great way to both sell everything that you grow and increase your price point per pound. But beware. Value-added crops take time and an additional investment in equipment that might be better used elsewhere on your farm. On the other hand, sometimes farms will discover a value-added niche that shifts the entire focus of the farm.

Steve Lecklider at Lehman's Orchard explained that he does everything he can to add value to his leftovers. He turns apples and pears into ciders and wines. He dries leftover cherries, apricots, blueberries, and apples. He has started making pure vinegars, including apple cider vinegar and red and white wine vinegars. His foray into value-added products has changed the direction of his farm. Instead of just selling fresh fruits at Chicago farmers' markets, he now sells wines through a distributor and is expanding sales through his new retail shop in town.

Another way to value-add is to host meals. Farm meals can be a good way to turn extra produce into profit. Many farms plan for meals around peak harvest season, when there are likely to be extras, as a way to add value for customers. We partner with a chef to host occasional farm-to-table meals in our barn. We start each meal with a tour of the farm, so we can show guests where ingredients for their meal were grown and tell them about growing the food. Food tastes better when customers know the story behind it. The meals can use up food in times of abundance and give customers a closer connection to a local farm.

Lehman's Orchard owner Steve Lecklider with wines and vinegars made from leftover fruit.

Meal in our barn loft.

Develop Markets for Leftovers

Some costs of overproduction can be recouped by selling in bulk. This does not mean price-slashing an hour before the market closes in order to clear your table. That practice merely diminishes the value of your booth, undercuts other farmers, and accustoms customers to wait around for cheap food rather than paying a fair price.

Rather, it means creatively finding new customers outside of your retail out-lets. For example, sometimes we advertise bulk quantities on Facebook for same-day farm pickup. Or we sell tomatoes in 5-gallon buckets to Amish neighbors to preserve. It might mean preserving the crop on your own to resell. A fruit farmer at our market freezes extra strawberries and raspberries to sell in bulk through-out the winter. Whatever markets you develop, the goal is to turn everything you grow into cash, as long as the costs to value-add return a sufficient margin.

required to yield enough carrots to supply our markets during all four seasons. Charts and graphs in seed catalogues and books aid in this, but the best data is our own. We keep a spreadsheet that tells us yield per bed for our crops and how many transplants fill a bed. For example, I know that a 60-foot-long bed of hybrid tomatoes in the greenhouse will produce approximately 900 pounds of tomatoes and require forty transplants. In addition, the bed will consume a half a square yard of compost. Year-to-year field maps help us adjust plantings as needed. For densely seeded crops, we often refer to seed purchases from previous years to help us get quantities right.

3. *Track seeding and harvest dates to time output with projected sales.* For a vegetable farm, projecting time to harvest for summer crops isn't hard, since seed companies often write the days until harvest right on the seed packet. But for winter sales we needed more information. Winter days are shorter, so crop cycles are much longer than in the summer, and they vary widely. We spent two seasons keeping tabs on a calendar when our winter crops went in versus when we harvested them. We put the information into a spreadsheet that we use to guide our planting to this day (see Table 10.2 on page 181). We found that some crops can take four weeks from seed to harvest during late fall but require more than four months in midwinter. Understanding precise seeding times has cut our overproduction way back, and we are able to target crops to reach peak harvest when prices are highest.

Forecasting helps produce the right amount. It's never possible to get it just right—markets always change and weather alters yields—but it is possible to come closer and closer every year.

HARVEST PRECISELY WHAT YOU PLAN TO SELL

Just as important as careful forecasting is precise harvesting. With vegetables, it's best to put together an accurate list of items you need to fill orders, then harvest precisely to the list. This is more efficient than harvesting first and scrounging for markets second.

We accomplish this by calling or texting all of our wholesale accounts a day or two ahead of harvest to get their precise orders. We add to those figures our projection of farmers' market sales and what we plan to put in the CSA boxes. We tally the numbers and write them on a whiteboard in the middle of our processing area for everyone to see. On harvest day we pick exactly to the list.

Ewert at Bair Lane Farm told me he also takes the principle seriously. He said it can be difficult to harvest bulk items, like greens, in precise amounts because it's hard to judge in the field exactly how many pounds he is picking into a tote. A tote of baby romaine might weigh twice as much as a tote of curly baby lettuce. To solve the problem, he harvests greens at the same time that his crew is washing them. As he harvests, he takes greens up to the processing room to be washed. He tries to match his pace to his washers so he never overpicks: "If we need 100 bags, I'll wait until they wash 99 bags, then run out and pick one more." For many farmers, more precise harvesting like this could reduce an enormous amount of waste.

Tool 2: Cut Costs to Grow Profit Margins

Business success is often measured by sales volume. To be sure, without paying customers there is no cash with which to keep the doors open. Almost by instinct, many farmers assume that the only way to grow their businesses (and profits) is to sell more. But cutting costs is an equally legitimate way to grow and comes with several benefits.

Let's look at an example. Say your goal is to increase net profit by 50 percent over ten years. What farm wouldn't love to do that! You could decide to increase sales volume. That would require more staff, more land under cultivation, more greenhouses, bigger processing areas, more delivery vehicles—an infusion of investment and effort.

Another way would be to cut costs. If you manage to trim your expense ledger by just 5 percent a year over ten years while keeping revenues exactly the same year to year, that would also grow your business by 50 percent, without extra effort or investment. The result would be a much leaner farm that is probably more pleasant for you and your staff to work on.

The above example is simplistic. In all likelihood your business will grow in a variety of ways, including through revenue increases and cost-cutting. Lean thinking simply encourages a focus on cost-cutting growth.

In fact, lean says that cost-cutting growth is the most efficient way to grow. Let me explain by looking at a simple equation:

$$sales - costs = profit$$

Let's say sales equals 10 units and costs equals 5 units. Profit would be 5:

$$10 \text{ (sales)} - 5 \text{ (costs)} = 5 \text{ (profit)}$$

You can expand profits by adjusting either sales or costs. Lean focuses on eliminating costs:

$$10 \text{ (sales)} - 3 \text{ (costs)} = 7 \text{ (profit)}$$

Decreasing costs by two digits increases net profit by two digits. It's that simple. Lean argues that for the effort involved, decreasing costs is a more efficient way to grow than focusing on increasing sales because increasing your supply is not cost-free. There is no way to increase sales without having costs go up as well. Not only will you have the costs of doing business (called variable costs), but you will also have all the wastes that scaling up is bound to produce, such as construction messes and interruptions in production. Hour by hour, your efforts will yield less total profit than simply cutting costs.

Let's look at some examples. In 2015 the fictional Happy Meadows Pineapple Farm is ten years old and generating annual sales of $100,000 per year operating at full capacity. Their expenses are $50,000 per year, leaving a net of $50,000 (a gross-to-net ratio of 2:1). Sally and Harry, the owners, would like to grow their profits in the next five years by 50 percent, so that their net by 2020 will be $75,000. They have two options:

1. Grow by getting bigger: add on to the processing room, plant more pineapples, find more customers, hire more staff.
2. Grow by getting leaner: root out wastes and cut costs by $5,000 per year, adding $25,000 in net profits over five years.

They decide that they are doing everything possible to grow pineapples as efficiently as they can, so they choose the growth-through-sales model

Value: More Important than Cost

Cutting costs is an important lean practice, but cost-cutting is not an end in itself. Lean production is not about racing to the bottom to create cheaper and cheaper goods using the fastest method possible. It's about recognizing what matters to customers and then refocusing efforts by eliminating projects, tasks, tools, and whatever else doesn't add value to your product.

On our farm we take waste elimination seriously, but in strategic ways we sometimes forgo less costly options in favor of higher value for our customers.

For example, we package our microgreens in more expensive plastic clamshell containers rather than cheaper plastic bags because clamshells are easier for chefs and market customers to handle without damaging the greens. Cut costs, but not to the point of decreasing value.

(the first option above). After five years of pushing production and sales, they reach their goal and are grossing $150,000 per year and netting $75,000. Because their process did not change, they have the same 2:1 gross-to-net ratio. But to get there they had to wade through a pile of waste. They had to spend time training new workers, cleaning up a big construction mess, haggling with contractors, and on and on. They put in lots of overtime and experienced plenty of stress caused by the expansion.

Their neighbors, Judy and Bill at the similarly fictional Happy Hill Pineapple Farm, reached a similar crossroads in 2015. They wanted to grow 50 percent more profitable by 2020. Their sales in 2015 were also $100,000, with expenses of $50,000. Instead of getting bigger, they wondered what would happen if they kept sales the same but trimmed costs every year by 5 percent, or $5,000 per year. Every year for the next five years, they print out an expense ledger and scour their operation for waste.

After five years they attain their goal as well. Their sales stayed the same ($100,000 per year), but their expenses were only $25,000, producing the same $75,000 in net profit as their neighbors but without the hassle. The key difference is that through leaning they enjoy working on their farm more—it's cleaner and work is more pleasant—and their business is able to withstand more ups and downs. In addition, they have freed up additional capacity (time and money) to add another crop (banana trees?) to their operation in the next five years if they so choose—or take more vacations. Eliminated waste leaves a vacuum of time and resources to fill any way you please.

▦ ▢ ▣

Here, then, are reasons I like lean (cost-cutting) growth:

1. The savings (and higher profits) are perennial with no more work. They add up. Imagine that you develop a way to wash vegetables that shaves twenty minutes a day off your workload. Over ten years (assuming a five-day workweek) you will have freed up 52,000 minutes—or thirty-six days—to grow more food or take a vacation. Or say you find a cheaper source of bedding for your animals, saving you $1,000 per year. Over ten years the savings multiplies to $10,000 in additional profit with no additional effort. The eliminated waste turns into capacity. It frees up time and money.

2. Getting rid of waste requires less overall effort than keeping your current process and increasing sales. More sales means more staff, more buildings, more supplies, more stress, more work.

Cutting Costs by Cutting Ice

My favorite illustration of cost-cutting is from an Amish farmer friend of mine, David Bontrager, who grows produce for Whole Foods. A few winters ago I visited David on a bitter-cold day in February. I walked up to the house, but no one came to the door. I looked around and didn't see anyone around the barns. Then I heard a small gas engine in the distance. I walked down a lane and saw a team of horses and a parked wooden hay wagon.

The wagon was perched on the edge of Bontrager's pond. The engine I had heard was an ice saw, a giant circular saw with a three- or four-foot blade. Bontrager was out cutting massive blocks of ice from his pond. As each block was cut, one of Bontrager's helpers would tie a thick braided rope around the ice chunk and guide it as the team of horses pulled it up a ramp and onto the wagon.

Ice for refrigeration was commonplace not long ago. Railroads hauled blocks of ice for people to use in iceboxes and coolers. But I had never seen ice extraction in action. Bontrager later told me his system is rustic compared to other Amish produce growers, such as those in Canada, who use ice refrigeration on a larger scale. According to Bontrager, those growers—with sheet metal chutes, elevators, and conveyers—can pull off two tons of ice in three minutes.

After the ice was loaded and stacked up like bales of hay onto the wagon, the horses pulled the wagon up to Bontrager's walk-in cooler, the box of a used delivery truck that he has insulated with foam. Bontrager and his help unloaded the ice, packing sawdust between each layer of blocks. Bontrager's cooler is a hybrid system, cooled both by a generator-powered compressor and his ice. His day on the ice shaves hundreds of dollars off his refrigeration costs. His ice should last at least until next winter. He explained that in coolers where floors have plenty of insulation ice could last for up to two years.

Besides saving him money, harvesting ice gets Bontrager out of the house and builds community during a time when he has a smaller physical workload and there is less happening on his farm.

Every dollar that Bontrager saves by cutting ice translates directly into a higher profit margin at the end of the year. With every block of ice that comes off his pond, he is growing his business without having to plow up more land and grow more food.

3. Eliminating waste makes your business more flexible, resilient, and pleasant for workers (including yourselves).

■ ■ ■

On our own (very real) farm, we have focused on waste reduction for four years, while our revenues have stayed about the same. We've managed to cut our costs from between 60 and 70 percent of gross to around 40 percent, thus increasing profits without ballooning production. We look for waste in the value stream as we work, and we print out an expense ledger every so often and ask, "Where's the 5 percent waste?" There was plenty of waste when we started, so trimming was easier then than it is now. But there is still plenty to identify. We've learned to enjoy the challenge almost as much as we enjoy growing food. Cutting costs is easier than expanding. And our farm is more pleasant to work on.

When asked about the need to calculate costs, Ohno famously replied, "Costs do not exist to be calculated. Costs exist to be eliminated." Ohno was so passionate in his distaste for costs that he wanted to spend as little time pondering them as possible: "I say you don't need cost knowledge. I'm not even interested in learning the terminology."[4] This is not to say Ohno was not *conscious* of costs. In fact, he made a career of finding costs and rooting them out. But once you identify a cost as waste, why bother to think about it? Just get rid of it!

Tool 3: Replace Low-Profit Items with High-Profit Items

One of the best ways to lean your farm is to spend time deciding what *not* to do. Many crops or animals are fun to care for but yield low returns. Take the energy you were spending on low-return projects and put it into producing more of your best-selling items. It might sound obvious, but it is surprisingly hard to do. "We have to produce X!" you might say. You can produce X, but realize it might be bogging down your farm.

Don't take the principle to an extreme, though. You can become a mono-crop farm in a hurry by determining your top moneymaker and stopping production on everything else. Like all lean tools, this one should be used in context—to support your farm and your family, not replace values like diversity.

When we started growing ten years ago, we planted and sold more than eighty different varieties of vegetables. By now we have cut that number in half. Each year we ask, "What were our top crops?" And then, "What would happen if we stopped growing the poorest-performing three or four crops?" I think of this as placing focus on winners—high-margin crops with strong local demand whose performance is consistently successful using organic techniques. We still need diversity, for the health of our land and to fill our customers' orders. But setting profit goals for each crop helps us focus.

A few crops that we stopped growing were ones that our customers valued. For example, we stopped growing winter squash and watermelons because they took up too much room (field management costs were too high with our setup) and because we couldn't move them efficiently. We were wearing ourselves out lifting them by hand, and we didn't want to remodel the barn to accommodate a pallet-moving system.

Just because we can't grow these crops profitably doesn't mean others can't. Other farms are set up to grow many crops more efficiently than we can. We partnered with a nearby Amish farm to provide those items in our boxes. We are up-front with our customers, who have told us they don't mind. Farm-to-farm partnerships often add value for customers.

Almost invariably, successful long-term farms stake out a niche—tomatoes, potatoes, garlic, greens; rabbits, sheep, hogs, heifers; apples, grapes, peaches—and develop special expertise over many years on their niche crop. In the beginning it might be appropriate to try to produce a little bit of everything, but as time goes on a farmer is wise to make careful decisions about where to focus a farm's energy.

Tool 4: Maximize Fixed Costs

Fixed costs are costs to your farm no matter what your level of production may be. They include payments for long-term investments like land, fences, tractors, greenhouses, and buildings. Variable costs are costs that rise and fall as production increases or declines. Think seeds, electricity bills, and gas. Lean production encourages a focus on maximizing production while keeping fixed costs the same. In other words, before buying new equipment or adding infrastructure, first use what you have to its fullest capacity (see figure 6.1).

From an accounting standpoint, lean practices like cutting costs, producing only what sells, and replacing low-profit crops with high-profit crops are management decisions that maximize fixed costs. They make the most of your farm.

Toyota was famous for its tight use of space for a firm of its size. When times are good, it is tempting to spread out, buy big equipment or build bigger buildings, and make life easier. Instead, Toyota kept economizing, choosing to maximize fixed costs rather than sprawl out. Womack and Jones write about touring Toyota's Takaoka plant in Toyota City, where they were astonished at the plant's narrow aisles:

> *Toyota believes in having as little space as possible so that face-to-face communication among workers is easier, and there is no room to store inventories. GM, by contrast, has believed that extra space is necessary to work on vehicles*

A quick crop of beets and green onions grows in the aisle between our tomatoes. In similar fashion, strawberries grow in unused aisle space between Riesling grapes at Lehman's Orchard. The practices maximize fixed costs.

needing repairs and to store the large inventories needed to ensure smooth production.[5]

The authors noted that because workflow was smooth, worker tasks were balanced, "so that every worker worked at about the same pace."[6] Spaces that are used to their fullest capacity are also efficient workspaces, reducing motion waste.

We used the "maximized fixed cost" principle when building our four greenhouses. We didn't build them all at once. We built one greenhouse and told ourselves we couldn't build another until we were selling every single item out of the first one. By the end of our first season, we were ready for a second greenhouse. We then applied the same rule: no third greenhouse until we were selling everything out of the first two. As of this writing, we have four greenhouses covering 9,000 square feet.

We also treat cultivated land as a fixed cost. This means growing as much value as possible on what we have tilled up before expanding to new ground. Often a new crop will go in within hours of the old crop's coming out. And we have spent years discovering ways to grow food in our farm's hidden spaces. Between our rows of young greenhouse tomato plants, for instance, we use aisle space to grow a quick crop of radishes or green onions or microgreens that will be harvested before the tomatoes are large enough to shade out the small crops.

A vertical garlic drying rack at Jericho Settlers Farm saves space.

Many farmers I've met have come up with other smart solutions to maximize their fixed costs. At his fruit orchard Steve Lecklider grows strawberries, tomatoes, and other spring and summer crops between rows of fruit trees. This is an unusual practice in orchards, where trees are often spread out and aisle spaces are wide and usually seeded in grass. Lecklider sees large aisles to mow as a form of waste: "If I can grow something there, then why

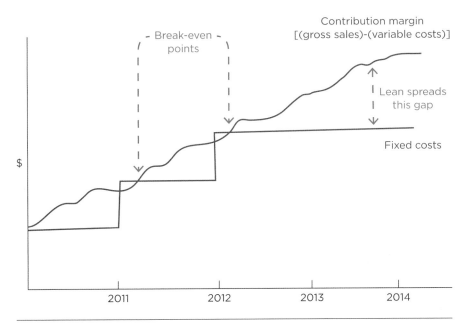

Figure 6.1. Lean spreads the gap between fixed costs and your profits, shown here as the contribution margin. In the early years, paying for fixed costs like tractors and land can mean slim profits. Maximize fixed costs, however, and your profits will grow. Graph courtesy of Steve Brenneman.

not?" He told me he uses the understory crops as a way to pay for the costs of growing the fruit crops. In 2014 he sold $30,000 worth of strawberries grown in the unused aisle space between trees. Whereas most orchardists must wait until midsummer or fall to collect a paycheck, Lecklider did not: "It's great to start the year off with an infusion of capital."

Christa Alexander at Jericho Settlers Farm in Jericho Center, Vermont, showed me a new vertical garlic-drying rack she uses. After garlic is harvested, it needs plenty of air moving around it for several weeks so it can dry and cure. Many farmers hang it in bunches on rafters or spread it out on a barn floor, which can gobble up a lot of room. The rack Jericho Settlers uses consists of several 4′ × 4′ stackable racks, saving space in their processing area.

Pete Johnson at Pete's Greens in Craftsbury, Vermont, told me the heated concrete slab in his greenhouse is his best example of maximizing a fixed cost. The slab provides bottom heat to vegetable starts and much more. "The slab grows veggie starts, cures squash, dries onions, stores equipment, and grows winter shoots," Pete explained. "There are few weeks in the year that it is not full and fully utilized." The heated slab greatly increases the utility of his greenhouse, a fixed cost.

At Blue Heron Farm, Adam Derstine takes pride in minimizing the fencing—a fixed cost—that keeps his hogs in their pasture. He told me, "I use deep-cycle batteries, single-strand fence, and $100 fence chargers, and the pigs stay in. Other people use $600 fence chargers, three to five strands of wire, and an obscene number of fence posts. Their pigs also stay in."

Sandy and Paul Arnold at Pleasant Valley Organic Farm in Argyle, New York, have carefully chosen their equipment and greenhouse capacity to match their market demands and their labor capacity. They want to maximize their fixed costs. "If I have more greenhouse space or more cultivated land than I know what to do with, then I am working for my farm rather than letting it work for me," Paul explained to me. In recent years the Arnolds have scaled back from 7 acres to fewer than 5 in cultivated production, "and we've seen our sales go up, not down."

Tool 5: Level the Load (*Heijunka*)

I can remember picking tomatoes one summer night by headlamp until after midnight—not by choice, but because we were so overwhelmed with work that our workday had turned into a work night. We were often overburdened in summer. By contrast, we spent winters waiting for spring rather than producing.

Through load leveling (*heijunka*)—the practice of spreading work and sales out as evenly as possible throughout the course of a day, week, or even a year—we now consider ourselves a four-season farm. Our area of biggest growth is in the winter. Our load isn't perfectly level, but it's much more even than it used to be, resulting in a more even pace and more predictable work.

An uneven load is full of waste. We can chart the two different ways to increase sales by 5 percent over a period of ten years (see figure 6.2).

In the first line, sales are uneven, causing a company to quickly scale up to increase capacity, then downsize, then scale up, then downsize, and so on. While total sales do increase, erratic production means enormous wastes:

- Staffing wastes because of the extra time it takes to constantly hire, lay off, and rehire workers
- Waiting wastes because of construction lag times
- Equipment burden from overuse during rapid increase
- Wasted moves and overburden of workers because of a chaotic work environment

In the second line, sales increase steadily; production is even. The load is close to level. This has several benefits:

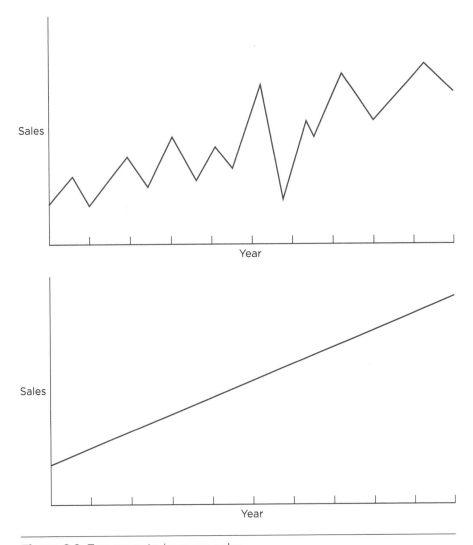

Figure 6.2. Two ways to increase sales

- Workers are not asked to rapidly scale up beyond capacity in boom-and-bust fashion, minimizing overburden on people and equipment.
- Work is more efficient. Process waste is easier to see because the work environment is predictable and smooth. As a result, defect rates are lower.
- The work environment is more pleasant for workers because work is more consistent and controlled.

Farmwork is by definition seasonal, not steady. At harvest time many farmers simply need a lot of hands to perform a lot of work. Still, farmers are wise to do what they can to keep work predictable and farm growth steady, not erratic. Any farm can find ways to spread out work by stretching

A Trip to Level the Load

When farms in northern places level their workloads, customers around them benefit: they can eat local all year. But replacing apples from Chile and carrots from California is no small task, especially in the winter. According to author and food journalist Michael Pollan, how to produce enough food to feed our communities in the winter months is "the big question facing the local food movement."[7]

On our farm over the course of seven years, we've developed techniques for keeping alive spinach, lettuces, kale, carrots, and more through the coldest, darkest parts of the year. Still, we're at the tip of an iceberg. There are hundreds of other varieties from all over the world that we've never tried and many growing techniques we've never explored.

Fortunately, other growers love the winter challenge, too. One day last August, I joined around twenty of them to compare field notes on what has worked

and what hasn't to keep things growing in the winter. I saw the conference as a way to learn how to level out our farm's load.

I was picked up at the Burlington, Vermont, airport by Christa Alexander, one of the owners of Jericho Settlers Farm. After helping her team unpack from a day of selling at the farmers' market, I spent the next two days in a room with other growers as well as representatives from the University of Vermont and Cornell University plus a few seed companies.

Fifteen years ago, commercially grown greens and vegetables in winter months in the north were virtually unheard of. Now there is a body of experience to learn from. Growers shared their experiences with a promising range of alternative and conventional heat sources (in-ground heat through buried water tubes, air heat from gas, convection heat using biomass boilers), new agricultural fabrics and plastics (greenhouses within

production seasons and sales and by performing as many tasks as possible during slower seasons.

Pete Johnson at Pete's Greens in northern Vermont has forged new winter markets for his greens and other vegetables. He told me that through storage and winter production, "our smallest week of sales in the winter is now two-thirds of our biggest summer week of sales." This is an incredible feat in a climate that often provides just two months of frost-free weather. In similar fashion, Steve Lecklider at Lehman's Orchard saves the winter months for value-adding his leftover fruit and selling value-added products. His winter sales are a significant part of his bottom line.

We appreciate the slower pace of winter—there is something to be said for time off and seasonal rest—but because we also realize the benefits of a more level work pace, we keep the farm going all year, though not in full a greenhouse), and countless new seed varieties—spinach, lettuce, Chinese cabbage, winter-hardy kale, komatsuna, and more—developed for winter production. Growers also discussed tips for improving storage practices in order to keep fall crops like carrots, sweet potatoes, leeks, and butternut squash around until spring.

Load leveling requires more than meeting production challenges, however; it also requires new market development. It's possible—even fun—to find winter customers. We started offering a winter greens share to our CSA customers in November and December. It's a box of salad mix, spinach, and many kinds of Asian greens that we package as a stir-fry mix. We also include carrots, salad turnips, and other items customers can add to their fresh salads. At our farmers' market we've expanded our range of storage crops like leeks, sweet potatoes, and winter squash, and we've found ways to stretch the growing season on fresh crops like ginger, microgreens, and

head lettuce so that our booth stays full year-round. In his essay "The Pleasures of Eating," Wendell Berry says "eating is an agricultural act"—what we eat determines to a large extent how land is used. The more customers put winter greens in their mouths in January and February, the more motivated we are to make our land productive at that time of year.

At the end of the conference, Christa Alexander dropped me off at the airport about an hour before my flight, enough time to unwind and grab a bite to eat. I stopped at a local favorite, the Skinny Pancake. Actually, it was the only restaurant in the terminal, and it hopes to become "the most local airport eatery in modern history." On the overhead menu-map, I saw Alexander's farm listed, as well as Pete's Greens and other Vermont farms, along with a big sign boasting: "Where good food comes full circle." I imagined eaters at the bar in mid-January, perhaps unaware that their dining choice was helping these farmers level their loads.

gear during midwinter. We end our CSA in December but continue delivering to a farmers' market all winter. In addition, we create a winter projects list. We want to do as much as possible in the winter to prepare for the busier summer. Our list includes tasks like oiling tool handles, sharpening blades, decluttering, and fixing tools. Bed preparation is also on the list. In early winter, when the ground is still dry enough to work, we prep our beds—chisel plow, till, shape, or add compost—in order to jump-start our spring planting.

Load leveling requires both raising low points as well as bringing down high points. For us, this means finding ways to temper our summer pace to reduce overburden. We've found that filling time in winter is easy compared to slowing down in summer. With time, we are getting better at scheduling in summer breaks, such as by taking off a few weeks in the middle of the

CSA season and keeping summer production in check. The goal is to keep our energy reserves stocked so that our work flows predictably and smoothly any time of year.

Tool 6: Use Metrics to Measure Your Work

Metrics are goalposts to assess progress on waste reduction in your operation. Lean managers use them to motivate workers and steer their operations. The most useful metrics follow these rules:

1. *They must be measurable.* If you can't measure a metric, there is no point in having it.
2. *They must be attainable.* Metrics, if used as goals, should be within reach in a reasonable amount of time. In Chapter 1 I discussed the idea that people are happiest when they experience flow, a state of deep concentration. One of the conditions for a flow task is that its level of difficulty should be *somewhat* challenging. If the task is too challenging, flow is broken and focus is lost due to frustration. If the task is not challenging enough, boredom breaks flow. In the same way, attainable metrics keep staff and farmers motivated (and happier) because they provide an attainable challenge and focus.
3. *They must be meaningful.* Metrics let farmers steer their farms in the direction of their visions. Metrics should be aligned with long-term vision. For staff, metrics should challenge workers to improve work in ways that are meaningful for them. Just because a farmer envisions making more money does not mean the metric will excite workers. Meaningful staff metrics can include goals related to waste reduction and quality work, and they are ideally developed by farmers and workers together.
4. *They must be simple to track.* The best metrics are part of accounting procedures you're already using or that can be assessed rapidly in just a few minutes per week. I spend very little time tracking what we do— maybe twenty minutes a week. I've

A lean planning meeting at Abildhort Dairy in Holstebro, Denmark. Ideally, workers should help create and track progress on lean goals. Photo by Susanne Pejstrup, Lean Farming®.

developed methods to track work in quick spurts on a continuous basis. Sometimes we simply check goals, such as yield per square foot, by calculating in our heads. I will quickly add up the square-foot value of crops as we're harvesting, then track the number once I get to the processing room. These are simple steps that keep production on track and that steer our farm in the direction of our vision.

Susanne Pejstrup, the Danish lean farm consultant, shared with me an example of metrics in use on a hog farm in Scandinavia. The farmer tracks:

- The number of pigs per sow
- The variation of weight of pigs for sale
- Food consumption per pig

The metrics are measurable, meaningful, and easy to track. Alongside each metric, the farmer sets attainable goals, such as 5 percent less feed consumption per pig. These types of goals keep her entire farm crew focused on cutting waste.

The Arnolds at Pleasant Valley Farm keep tabs on the dollar value per square foot of their crops, but they extrapolate out to dollar value per acre. If a crop makes the cut for them, then it must produce a set amount per acre. The metric helps them steer production in the direction they want to take their farm.

In a similar way, Jack Algiere, farm director at the Stone Barns Center for Food and Agriculture in Pocantico, New York, told me his crops must pay a set amount of rent per plot. One year he projected that a plot of squash could not fully pay the rent for an area, so he planted a few rows of cucumbers next to the squash to bring the value of the area up to his goal.

If a crop cannot pay the rent, that doesn't necessarily mean you should eliminate the crop. It could mean you shouldn't grow it at that time in that place. On our farm head lettuces do not pay their way in a January heated greenhouse because they grow too slowly, but they will be worth growing later in the spring, in an outside plot, where costs are lower.

Algiere knows the rent his crops need to pay at different places on the farm in different seasons. He can use this information to charge chefs for experimental crops. He might tell a chef that he'll agree to grow a cubic yard of snow peas in a heated winter greenhouse, but he will charge a set amount for whatever yield he gets, because that is the rent for the area. "That way the farmer isn't always the only party taking a risk," he explained.

Lean Recordkeeping

Metrics and other tools for long-range farm planning might be important, but keeping track of metrics is still a form of type 1 *muda*, since the actions add no direct value to your product. Minimum is the goal. I recommend three principles for lean recordkeeping and planning:

1. *Apply 5S to the office.* Get rid of items cluttering your desk and delete from your computer files that don't serve a farm purpose. If you can afford it, dedicate a computer for farm tasks to keep computer time focused. Don't be a catalog collector. Keep only the ones you need. Remember, everything you keep is a cost.

2. *Keep data-tracking materials close at hand.* My favorite example of a simple data-tracking system is from Randy Ewert at Bair Lane Farm. He keeps two calendars next to his seeds, last year's and this year's. When he plants, he looks at last year's calendar for notes he might have written to himself, such as "Plant four beds of broccoli next year" or "Cover salad turnips with row covers right away." After he plants, he writes in this year's calendar how many row feet or beds went in. When he harvests, he writes down whether he planted too much or too little and other notes for next year. It is a very simple, quick way to pass data on from one year to the next.

On our farm, we keep data tracking efficient in a couple of ways. We put sticky notes on each tote of market-ready vegetables in our cooler, indicating what the item is, the quantity, the day it was picked, and the destination. We keep markers and pads of sticky notes handy—one set in the processing room, another beside the cooler. I keep track of orders and weekly sales on a scrap of paper in my

Tool 7: Balance Creativity and Discipline: The 15 Percent Rule

Lean production is like two sides of a coin. On one side is creativity needed to perennially find improvements, to do things differently, and to test hypotheses. On the other side is the discipline of implementation, setting limits, defining the boundaries of your business. There is a yin-yang balance.

In the early years of a farm, experiments and creativity have an important role to play in testing crops and trying out markets to find a profitable fit. Often, though, innovation carries on for too long. Eventually, experiments need to be put in their place: failed systems need to be completely eliminated and the most profitable systems scaled up and further improved.

Recordkeeping calendars and field maps hang at eye level in our seedling greenhouse.

Totes stacked up and labeled in our cooler. Labeling supplies sit just outside the door.

pocket or my iPhone. I keep track of our metrics on a sheet of paper that I keep in the washing area. When I bring crops in, I walk right by the sheet. We hang our recordkeeping calendars and field maps on our germination chamber in the seedling greenhouse. There is no hunting around for them: they are at eye level, right where we need them.

3. *Don't bother tracking everything.* We learned early on that farming hours would be limited if we didn't prioritize the kinds of records we kept. We ask, "Which records help us eliminate waste, and which just give us interesting information?" We stick with the ones that help get rid of waste. These include records of seeding and harvest dates when we are testing new crops, records of amounts harvested on the Google Drive spreadsheet to project sales (Tool 1 above), and very simple plot maps to tell us how many beds of each crop went in each season.

If you've decided a crop or animal project is successful, stick with it. A farm that is sustainable for the long haul will need to transition from an innovation center to a production center.

In order not to get bogged down by experiments, I suggest a 15 percent rule after the second or third year of your farm: 15 percent of your time should be reserved for innovating systems to try out new ideas that are exciting to you (creativity), while the rest of your time should be focused on production systems that are known variables (discipline). Anything less than 15 percent and a farmer is prone to boredom or burnout through repetition, and your farm will miss out on new ideas for improvements. Any more than 15 percent and your farm could struggle because your time is gobbled up by your experiments.

We don't literally track every minute of our time to see how much we spend on experiments versus known variables. But we do try to limit the new items we grow every year to less than 15 percent of our total offering. We try to improve processes incrementally, in two or three new ways every year, while relying mostly on what we know works. For instance, new crops in 2012 and 2013 for us were figs, ginger, and turmeric. In 2014 we added oyster mushrooms and several new winter Asian greens. Process improvements in 2014 included a carrot digger, a second germinating chamber, and a water heater in the processing room. In 2015 we are growing new ginger and turmeric varieties. Our current focuses in process improvement are the portable roots washer and a Japanese paper pot transplanter.

Of course if you want a research farm and can find financing for inventions and innovations, then go for it. From a lean point of view, if your customers, granting institutions, or donors value a constant drumbeat of new

Big Kahuna ginger, one of our experimental 15 percent crops in 2012, is now a regular item in our rotation.

On the Benefit of Limits

The 15 percent rule might feel stifling, but there is a benefit to setting limits and sticking with them: the practice can unleash creativity.

Ohno tells the story of a government scheme in Japan to reduce rice production. The government mandated that farmers stop using 10 percent of their rice paddy acreage:

> The farmers were expected to produce 10 percent less rice, but the productivity of the remaining acreage actually increased by 10 percent and the actual production of rice did not change. [Government officials] said, "We did not reduce enough cultivated acreage," but they must have used pure mathematical calculations and ignored the idea of productivity. . . . They should have instructed farms to reduce their output by 10 percent, regardless of what acreage they used to produce it.[8]

The creative farmers found ways to tuck rice plants into unused corners of their properties and increased their productivity by 10 percent, confounding the government regulators. Constraints preceded innovation.

Thoreau plumbed the depths of Walden Pond—and his spiritual life—by staying put in one place, at least for the two years he spent writing *Walden*. He admonished his readers to limit their affairs to "two or three." With limits come depth, creativity, innovation, and a tendency to nurture what is within your reach. With unlimited boundaries come shallowness, lack of focus, and poor management.

An important way we've cut our costs is to find ways to produce the same amount on a smaller footprint every year. We started on 3 acres, then scaled down to two, and now we are farming less than one. All the while we've managed to keep sales steady—in fact, like the Arnolds, sales went up slightly each time we downsized the square footage we farmed, because the smaller footprint received better care and produced more. We save on inputs like composts, and we are walking a lot less, saving moves.

I know we are more creative on 1 acre of land than we ever were on the larger plots we've farmed. One year we planted several hundred more tomato plants than we needed and spread them out over ½ acre of land. The yield per plant was low because we couldn't possibly tend all of the plants well. Now we constrain ourselves. We grow only eight short rows in our greenhouses. Because we are focused on a small number of plants and are innovative with our space, we produce far more than we ever did on ½ acre of land.

Creativity thrives with a bit of discipline.

ideas, then you can structure your work to give them that value. If your goal, however, is to make a living through the sale of your farm's products, then don't get stuck at the wrong end of the creativity-discipline continuum.

There is still a place for quality of life. Hobbies and homestead projects—like experiments—might need to be put in their place, but there is no reason to eliminate them altogether in the name of lean. A farm is a home, not just a business, and quality of life matters. Lean is a tool—it can shore up the business side of your farming, but it doesn't need to strip what matters from you. Even though we don't sell animals anymore, we keep them around for our own sake. We love our own fresh pork, and we can't buy the quality of eggs we want. Likewise, I enjoy making our own pottery dishes out of our clay soil, and Rachel enjoys metalsmithing and jewelry making. These projects aren't money-driven; they're not lean in the sense that our time and energy might be more economically spent on other endeavors. But by applying lean in our production space, we *have* time and energy for these other pursuits.

■ ■ ■

Wastes that originate in the management office—decision wastes—are every bit as destructive as production wastes. These aren't low-hanging fruits; they aren't always easy to see. They might be conceptual in origin, but they lead to practical problems on the farm. Fortunately, with lean tools and a bit of discipline, you can get rid of them.

CHAPTER 7

Lean Farm Sales: Establish Pull, Don't Push

*The question is whether or not the product is of value
to the buyer. If a high price is set because of the
manufacturer's cost, consumers will simply turn away.*

—*TAIICHI OHNO*

My wife and I joined two friends a few years ago on a vacation to an island off the coast of Belize. The island was tiny, just 9 acres, and remote. The trip to the island took two and a half hours by catamaran. We were dropped off on Sunday, along with thirty other travelers, and told we would be picked up the following Saturday. Fees for use of the island were small because accommodations were simple. We slept in rustic huts with palm-leaf roofs. The huts sat on stilts over a reef, several yards offshore. No modern plumbing, no big-screen TV.

There was also no grocery store. To eat, we relied on coconuts and fish. We had brought along enough rice for a week. A problem was that none of us had ever fished in the deep sea. Casting a rod from the banks of Buck Creek in rural LaGrange County, Indiana, where I grew up, was nothing like pulling a fish from the Caribbean Sea.

Enter a quiet and cool-mannered young man I'll call Roman. Far as we could tell, Roman had grown up on the island (his parents were the owners) and made his living by offering fishing help to travelers like us, hungry vacationers who needed him. He had found a niche market.

Roman took my friend and me out on his boat early in the week. After about a half hour of trolling, letting the line drag behind the boat, Roman grunted. Because he rarely spoke, we knew something was up. A minute later he reeled in a small jack, a common fish in that area, which he used as bait to reel in a much larger barracuda a few minutes later. We had our main

Late fall in a greenhouse. Precision, not volume, increases value and pull.

dish. We took the fish back to shore, grabbed a few coconuts to grind into pulp for a sweet side dish, cooked our rice, and dined all evening.

Our eating on the island was a weeklong example of pull. Nothing was produced and then pushed on us. We didn't overstock. For every meal, we responded to our hunger and ordered fish and gathered together everything else we needed precisely when we needed it in the amounts we desired.

Pull Thinking: A Brief History

In its simplest form, pull selling is replenishment. Vending machines are a perfect example. Consumers pull candy bars; producers replenish candy bars. A vending machine in this sense is very lean. On the island we ate while nature (and Roman) replenished.

Ohno was inspired to institute pull systems at Toyota after seeing pictures in the 1950s of a modern US grocery store:

> *Back in 1951 or 1952, the first of our classmates to go to the United States came back with all sorts of color photographs that he took, the type that you display with a slide projector. Among them were several photographs of a supermarket. He explained that in the United States there was something called a supermarket, and there was only a young woman at*

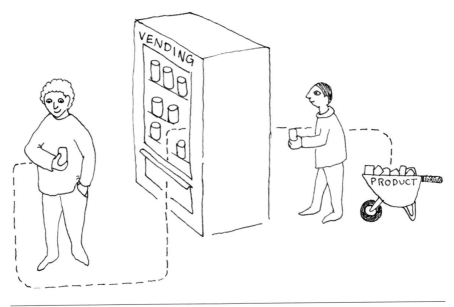

Lean Pull: Consumer uses products. Provider replenishes products.

*the exit, and the customers pulled along something like a baby
stroller, bought just what they wanted, and paid at the exit.[1]*

One woman selling all that food! The concept was novel. Japanese grocery sales at the time involved home delivery and "grocery men walking their neighborhood routes": "For example, even with tofu, in the morning when the tofu was ready the tofu seller would walk around playing a flute, selling tofu."[2] The system involved a lot of walking—a lot of people—to sell relatively little food. And consumers rarely got just what they wanted. Ohno explains: "If you order home delivery and you only need two stalks of long onion, you can hardly ask them to bring two stalks of long onion, so you order a whole bunch. You think, 'I might as well buy some daikon too,' so in the end this is an uneconomical way of shopping."[3] By contrast, in the American stores, "the buyers can buy according to the size of their refrigerators and amount of money in their wallets, and live economically."[4]

What made the stores so novel—so efficient—was the practice of *replenishment*. When shelves emptied, they were restocked. "Parts" were pulled by the consumer and replenished by the provider as they were ordered.

Toyota got to work instituting the same concept on the assembly line. Before, manufacturers might make 10,000 parts, then production workers would assemble 10,000 parts together into goods, and then sales would try to sell (push) the goods after they were made. With pull, the order is

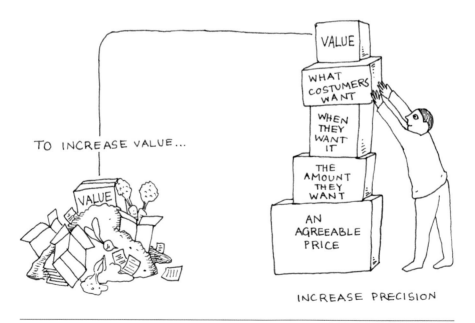

Sell farm goods faster by increasing their value, not dropping their price.

reversed. Customers start by ordering (pulling), and production workers pull parts as needed from manufacturers. "If you do this well," says Ohno, "increasing productivity about three-fold is easy."[5] The challenge switched from how to make large batches as efficiently as possible to how to communicate precise needs from one department to the next.

In a farm context, your goal as a seller of food is to have available what your customers want when they want it. Let them pull. Your goal as a producer is to replenish supplies as soon as they are pulled.

In short, the key to selling lean is to start way before you pull your delivery van into your farmers' market. Start by producing what customers want and planning ahead to supply it when they want it in the amounts they desire. Be precise. Precision, not volume, increases value and pull and lets you charge living-wage prices. This practice, coupled with lowering production costs on our farm, is how we achieve profit margins that sustain our family.

The Difference between Pushing and Educating

In the case of local and specialty foods, educating customers about your product is different from pushing. Last winter I led a farm tour for a man and his wife from India traveling through the area. They were astonished to see so many different kinds of food in such a small area and so many varieties they had never heard of. When I took them to our microgreens greenhouse, they couldn't believe we could sell crops so small.

Microgreens—a crop many enjoy once they taste it.

"Surely this must be a hard sell because most people don't even recognize what this is," the man told me.

"Try a pinch," I replied.

They sampled a few leaves of pea shoots, their eyes lit up, and immediately the husband said, "Now I see. Just let them eat a few leaves!"

Pushing is growing a lot of what *you* want to grow, then dumping it for whatever price you can get on the market.

Educating is giving potential customers a taste of a product they would otherwise never know existed.

Prices That Pull, Prices That Push

Ohno describes a wrong way to set price:[6]

$$selling\ price = profit + actual\ cost$$

In this equation, managers add up their costs, tack on a margin of profit, then establish a price. Let's say you are a milk producer, and your production costs are $1.50 per gallon of milk. You want to make a profit, so you charge $2 per gallon.

Farmers commonly set prices this way, but Ohno says it is wrong because "we make the consumer responsible for every cost." Instead, farmers, not their customers, should be responsible for their costs. If your production system is full of waste and your costs are high, that's your problem. Likewise, if you are lean and can produce high volume at low cost, then you have earned your high profits.

Prices, then, should be amounts that you and your customers agree to, based on external market factors, not internal production factors. According to Ohno, "Our products are scrutinized by cool-headed consumers in free, competitive markets where the manufacturing cost of a product is of no consequence." To use farming language, customers don't care about all the time you spent weeding or tending your animals; rather, "the question is whether or not the product is of value to the buyer."[7]

Don't waste hours calculating your cost to set prices. Set prices based on market realities. Charge what you can. If you've done your homework and created a product full of value, then standard market prices—agreeable prices—will give your product plenty of pull.

▪ ▪ ▪

Prices that push are prices that undercut the market. You've overproduced or perhaps your product doesn't match what customers value, so you flood the market and slash your prices—and your profits—in order to push the excess onto your customers.

Undercut. Slash. Flood. These are harsh terms. But the practice is harsh, too. For you as a producer, undercutting hurts your bottom line. Undercutting is the opposite of leaning. Rather than increasing value by creating a better product, you are "adding value"—making your product more desirable—by robbing your own profits. It's a shortcut, and it won't pay off. Sooner or later it will put you out of business. Undercutting also hurts other farmers. Why start a race to the bottom that will just make other hardworking producers suffer alongside you?

◻ ◻ ◻

A specific example on our farm is tomatoes. We love heirloom tomatoes. They often have more complex flavor than standard tomatoes, better texture, and with all their stripes and funky shapes, they look cool, too.

We also love to grow them. For sure, they're a challenge. They have odd growth habits. They send shoots in all kinds of directions and set fruits in unpredictable ways. They're prone to dozens of foliar and soil diseases. But their gangling habits and unpredictable ways also delight us.

When we first started, we grew mostly heirlooms. We had seen prices for heirlooms in urban markets and thought we could keep our costs low enough to grow them for a profit. It turns out that our customers preferred standard red tomatoes, which sell for less (and cost less to produce). The only way to sell a lot of heirlooms was to lower prices. So if we wanted to keep heirlooms on our farm, we had two options: grow lots of them and slash prices or grow a smaller amount and keep prices high.

We decided on the latter.

Heirloom tomatoes and red tomatoes.

Removing Overburden at the Farmers' Market

In the beginning we generally carted our food to market in bulk and sold a lot of items by the pound. We took in onions or tomatoes or beans in totes, set them in display baskets, and let customers pick up the amount they wanted, then charged them by weight.

That worked fine as long as markets were slow. But when sales were ticking, we would lose customers who didn't want to wait in line while we weighed up everyone's order. Weighing took too long and increased work because many orders needed to be added up with a calculator. The practice was a burden on staff. It usually took at least two staff people to handle markets: one person to weigh orders and another to take care of everything else. Because we handled food a lot (a lot of moves), we created a flurry of activity that sometimes led to charging people the wrong amount.

To streamline, we created flow.

We now sell in grab-and-go units whenever we can. We sell potatoes, onions, tomatoes, and peppers by the quart. When they are large, we sell those items by the piece, not by the pound, along with other big items like winter squashes and melons. We also quart up or bag beans, loose carrots, beets, green beans, and other smaller items. We sell figs and raspberries by the pint, and kale, radishes, and turnips by the bunch. The result is that customers can quickly grab what they want, head for checkout, and be on their way in a fraction of the time. There is no friction. To get ready for market, we package many items into their units on the farm earlier in the week so market setup isn't a mad rush.

To make the math even faster, we streamlined pricing so that almost all of our unit prices—no matter if the item is bagged, bunched, or boxed—are $2.50, $3, or $5. The only exceptions to the rule are items like pink ginger and sweet potatoes that are impossible to put into units because of their shapes. Added up, these practices have cut our booth staff costs in half. One person can now handle the job of two.

To increase flow at our booth we sell in grab-and-go units whenever possible.

Now we mostly grow standard tomato varieties, along with a few rows of heirlooms for two restaurants, our CSA, and those market customers who appreciate them. We are always adjusting the amount of both kinds of tomatoes that we grow based on the previous year's sales. Every year we also fine-tune our tomato production to remove more waste and lower our costs. We prefer these practices to lowering prices as a way to increase profit margins and pull on tomatoes.

Leaning the sales department, then, means loading your product with value, as if adding power to a magnet. Charge market prices. If your products are full of value, their pull will be strong and they will easily sell.

Lean-to addition to our processing area. The new space features roll-up curtains on three sides.

CHAPTER 8

Continuous Improvement
(*Kaizen*)

A foolish consistency is the hobgoblin of little minds.

—RALPH WALDO EMERSON

It is our human nature to get things wrong. We view the world from a subjective perspective, our own. Distortions cloud our understanding of what customers actually value, and distortions cloud how we see our production process.

Plato wrote that we live as though in a cave, and all we can see are reflections of truth on the wall, not actual truth. After years refining Toyota's production system, Ohno concluded, "People's ideas are unreliable things, and I would be impressed if we were right even half the time."[1]

Look at figure 8.1.

Pretend each dot represents one activity in your day, some positive events, others negative. Perhaps you wake up to a beautiful sunrise but can't find your toothbrush a few minutes later. Then you eat a pancake breakfast that turns your day around but get stuck in traffic on the way to work. On the job you receive a compliment from your supervisor, but your computer crashes, losing hours of progress. Later that night you enjoy a leisurely walk along the lake but then receive the difficult news that a friend is ill.

On the one hand, you could say you had a great day. On the other

Figure 8.1.

hand, you might say your day was a disaster. It depends on how you connect the dots.

Psychologists call this process "chaining," and a method of psychotherapy involves teaching clients how to read their dots in adaptive rather than maladaptive ways.

Now think about your farm. Pretend the dots are steps in your value chain. Some steps are incredibly efficient and add lots of value; others represent waste.

Let's say you grow lettuce. You start out with a germination procedure that gives you 100 percent germination, but your greenhouse is poorly ventilated, so only half the seeds survive to transplant stage. When transplant day arrives, you use a time-saving tool that both preps your bed and marks your rows, but your workers plant too deeply, causing some transplants to die. At harvest time you cut the heads that survived, using a single-piece flow process that shaves hours of work, but overly hot weather causes several heads to wilt before reaching the cooler.

Was the crop a success? The seeds germinated well, the starts were transplanted efficiently, and the lettuce was harvested and packaged quickly. Yes, success. But the crop also failed, since the greenhouse killed several

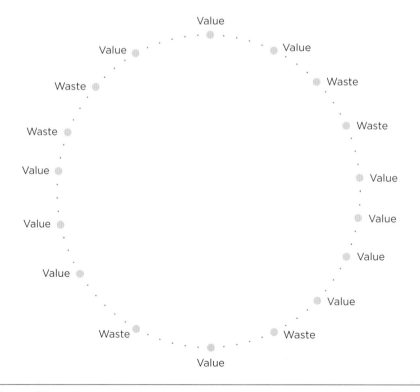

Figure 8.2.

starts, workers killed more when they transplanted them, and some lettuce wilted on the way to the cooler.

Lean asks farmers to continually analyze their dots. Unlike a psychotherapy session in which you might focus on the positive dots, lean wants you to look at "maladaptive" dots and not despair but be inspired to improve. The work of lean is circular, not linear. There is no final destination. The task of *kaizen* is to get more and more precise in identifying value, then to banish more and more waste so that your product jumps without pause from one dot to the next. Your goal, eventually, is perfect flow (zero waste). Perfection is a north star, a direction in which to head, even if it isn't always attainable.

Unless you are a one-person farm, there is no way you as a manager can see all the dots because you can't see all the way around the circle. Let's say on a typical workday you have a crew of four, including yourself. One person harvests carrots, another prunes the apple orchard, another moves the chickens, and you seed the greenhouse. From your dot you can perhaps keep a fleeting eye on one or two other dots, but unless you manage your farm with binoculars from a helicopter, there is no way to keep tabs on the whole system. That is why continuous improvement is a whole-staff exercise. You need many points of view to see reality. *Kaizen* and respect for people are linked.

In this chapter I discuss five tips for applying *kaizen* on a farm:

1. Develop improvement routines (*kata*).
2. Focus on the most-needed improvements.
3. Upgrade standards (which should be constantly changing).
4. Compare your farm to your farm.
5. Don't wait for things to go wrong to practice *kaizen*.

1. Develop Improvement Routines (*Kata*)

Continuous improvement is like a wheel perpetually spinning out better ideas. Ideally, your operation should rely on ingrained routines, called *kata*, for constantly evaluating your process and developing new ideas. The routines, such as regular staff meetings and daily farm walks, make improvement an integral part of everyday farmwork rather than a task set aside to fill up long winters (although winter is a fine time to improve systems). *Kata*, which originally referred to movement sequences in martial arts, should be practiced often so that they become as habitual as breathing, as planting seeds in the spring.

I think many creative farmers and farmworkers practice *kaizen* instinctively, in the same way that many farmers can see the flow of value well (see Chapter 2). These people habitually look for ways to improve. For others, however, implementing intentional daily or weekly routines for analyzing ways to improve is a useful discipline.

Rachel and I took most of the initiative to implement the first stages of our lean transformation. It took us several seasons to feel as if the farm had turned around and that the chaos I described in the introduction was under control. We used to perform most of our *kaizen* on the fly. I love field management and am always on the lookout for ways to perfect systems. If a tool is in my hand, I question whether it can be better or if I am using it with the most efficient motions. More than once I have interrupted my hoeing or transplanting to rush to the shop and grind a blade to a better angle or replace a short handle with a longer one.

Eventually, we found ways to routinize our process improvement. We start every week by walking around our farm with a pen and paper in hand. We make lists for the week and look for ways to improve what we're doing. We keep a notebook in our processing area with tabs for each vegetable. We track when we seed and when we harvest and jot notes for ways to improve systems for next year. I make it clear during our orientation and when we work that staff can help us by giving us ideas for doing things better. We use a whiteboard (see Chapter 9) to track our work and progress toward goals. There is still plenty of on-the-fly *kaizen*, but putting regular systems in place ensures that the wheel of new ideas is always going around.

It is important to solve problems. But it is more important to develop habits to keep addressing new problems as they arise. A farm that produces its product the same way over and over again will not thrive for long because new challenges, either on the farm or in the marketplace, are inevitable. The strength of your farm's *kaizen* routines—not a single solution to an isolated problem—will ultimately determine your farm's success.

2. Focus on the Most-Needed Improvements

Control the direction of your operation through priority-setting and then manage *kaizen* implementation so you make progress in logical steps. Work incrementally. Be strategic. Scour your value stream for waste. Then choose three or four most-needed initiatives to work on. Don't tackle the whole farm at once, though it can be tempting to try. Especially early on, we struggled to know where to put our energies. I can remember days going back and forth between building greenhouses, setting up wash stations, grading driveways, digging trenches, and buying equipment—on top of growing food. We didn't always manage our schedules wisely. However, occasionally we would back up and revisit our vision to make sure we were on track and to prioritize our work, choosing a few tasks to focus on. Without these sessions, we would have burned out.

The challenge with continuous improvement is to keep perspective and focus. Improving a farm bit by bit is like restoring a hundred-year-old house, a task we accomplished before we bought our farm: the project was manageable only through careful priority-setting, patience, and a long-range view. Lean farming calls for the same breakdown of the overall work of improvement into manageable pieces.

3. Upgrade Standards (Which Should Be Constantly Changing)

Much of the work of management boils down to developing and communicating standards, that is, routinized ways of getting jobs done in a consistent fashion.

Lean says standards should be fluid, not fixed. Ohno explains, "If you think of the standard as the best you can do, it's all over. The standard is only a baseline for doing more *kaizen*."[2] According to Ohno, when you discover better ways, "you should implement these ideas right away and make this the new standard." Ohno tells a story to illustrate the point:

> *Years ago, I made [team leaders] hang the standard work documents on the shop floor. After a year I said to a team leader, "The color of the paper has changed, which means you have been doing it the same way, so you have been a salary thief for the last year." I said, "What do you come to work to do each day? If you are observing every day you ought to be finding things you don't like and rewriting the standard immediately. Even if the document hanging here is from last month, this is wrong." At Toyota in the beginning we had the team leaders write down the dates on the standard work sheets when they hung them. This gave me a good reason to scold the team leaders, saying, "Have you been goofing off all month?"[3]*

It's easy to get into a rut on a farm, thinking you have discovered the best way to produce a crop. "The beans grew well last year, so why change what I'm

Standards should always change. We recently started raking all of our greens after harvesting as a way to improve second cuttings.

doing?" you might ask. There is always a way to improve. Each new improvement in process is not an end but a stepping-stone for further improvement. Removing waste is never a onetime task because waste morphs and shifts, just as a business changes shape over time. Getting rid of waste is a layered process. Only when one layer of waste is removed can you see the next. Also, new technology and tools are continually making new systems possible.

4. Compare Your Farm to Your Farm

Learn from other farmers as much as you can, but in the end go home and compare your farm to your farm. *Kaizen* is an internal discipline. It asks you to improve yourself based on your own problems and your own vision for your farm. The role of *kaizen* is to continuously improve the value stream that runs through your own farm. Perfection—the goal of *kaizen*—should be your own perfection. Your competitors can take care of themselves.

You might not have the newest technology or the fanciest tools. A lot of our equipment is farm-built and might work well only on our farm. But our methods work for our system, and we improve them every year based on our own setup. If you see an idea you like, test it out. Your goal, however, shouldn't be to try to catch up with or copy your neighbor but to improve flow in the unique context of your own farm.

5. Don't Wait for Things to Go Wrong to Practice *Kaizen*

The best time to practice *kaizen* is actually when crops are growing well or animals are healthy, because when things go wrong you are often fixing obvious problems, not truly improving. When times are good you have the most energy to stand back and get a clear picture of which activities can be done even better.

I remember coming home from town recently and finding a lake in our processing room. A plastic ball valve in the water line had cracked because of cold weather and filled the room with 4 inches of water. I ran to the house and shut the water main off, and Rachel and I bailed out water and cleaned up the room for the rest of the evening. In a hurry, I spliced the line with another plastic valve from the shop.

Later, when I had more time, I returned to the room, this time to rethink our plumbing, to really improve. I replaced old plastic water lines with newer PEX tubing, a polyethylene plumbing line that resists cracking. I installed a

Every farm is different. Your goal should be to improve flow in your own context.

new, bigger water heater and cut foam insulation for the windows to keep the room warmer. I replaced the plastic valve with a durable brass valve. Like the cracked water valve, things always seem to go wrong at the worst times, when we're busy. The challenge is to improve them when we're not.

▪ ▪ ▪

Kaizen is looking at the "bad" dots in your circle and focusing on them rather than ignoring them. The practice is at the heart of lean. I remember when we first applied the 5S organizing principles to our storage room. I thought, "This is it. We're done. We've leaned up the farm." But then we revisited the room six months later and removed just as many useless tools and junk as we did the first time.

And then I thought, "Okay, it took two rounds—now we're definitely done." But really, all the cleaning up was just step one. Rooting out process waste was step two. Rooting out management waste was step three. *Kaizen* is step four. And *Kaizen* never ends. It's tempting to think, "We've figured it out; now we can relax." But as Womack and Jones write, "When you've fixed something, fix it again."[4] This is *kaizen*.

CHAPTER 9

Respect for People: Lean and Farm Staff

People don't go to Toyota to "work"; they go there to "think."

—*Taiichi Ohno*

If the whole team is to perform *kaizen* well, you need well-trained, well-treated, up-to-speed staff. Respect for people goes hand in hand with *kaizen*. Managing staff must be done well because everyone involved must be able to see the flow of value and their role in creating value.

Toyota's success is due not just to its technical system focused on high-value flow. Just as important, according to Jeffrey Liker, professor of industrial and operations engineering at the University of Michigan, is Toyota's habit of nurturing people in "an organizational culture that expects and values their continuous improvements."[1]

Toyota managers build people, not just cars.

A misconception about lean is that its primary goal is to use cheaper and cheaper labor or to replace human labor entirely with machines. The opposite is truer. During the 2000s Toyota was rapidly building new production capacity in the United States while US manufacturers were closing plants and moving production overseas. Lean is not a race to the bottom in pursuit of less expensive labor, expecting workers to plug along, heads down, grinding away at meaningless jobs. In lean workplaces highly skilled workers carry real responsibilities and are actively engaged in making improvements.[2]

Lean Staff Management Tools

Lean staff management can help integrate workers on your farm by focusing on efficient training and ensuring that your standards are visible and unambiguously clear. Below, I discuss three techniques:

1. Training within industry (TWI)
2. Standard operating procedures (SOPs)
3. Visual system management (VSM)

I'll show how they've helped us, and I'll show them in use in a few other contexts as well. Then I'll discuss the lean concept of pushing responsibility down the organizational ladder.

1. TRAINING WITHIN INDUSTRY (TWI)

The goal of training within industry (TWI) is to ensure first-quality work the first time. The US government developed TWI during World War II to rapidly train thousands of workers entering factories to build war machinery. The tool can quickly get workers to a point where they can accomplish a task well, in a timely manner, and without supervision. We use TWI in three steps:

1. *Train yourself first.* The method starts by placing responsibility on the trainer. If a job wasn't done right, it was the trainer's fault, not the worker's. According to TWI, a supervisor has the responsibility to know the work well.

   Before I begin training, I make sure that I explore a task for potential problems. For instance, when I developed a new method for screening salad mix from one wash tank to another, I first used the screen on my own for several batches in order to work out kinks and to predict the hiccups a new worker might face.

   After I developed our system for growing microgreens, I typed out steps and then followed the steps on my own for several weeks before passing the assignment along. In the process I learned about potential problems, such as what happens when the growing medium is too wet or when a tray is seeded too thickly; I was able to pass on my newfound expertise to a worker. The point is to train what you actually know, not what you expect someone else to discover.

2. *Break work into tasks.* The next step is to break the sometimes complicated and interconnected work of farming into tasks that a worker can follow from start to finish without supervision. Our goal is to train new workers (except for day volunteers) on a few basic tasks within the first few weeks of their arrival. Here are some examples for us:

   - microgreens seeding
   - greens washing and packaging
   - roots harvesting and packing

- greenhouse seeding
- operating the spray station

3. *Break tasks into important steps, key points, and reasons.* The fastest way for a task to be learned is in steps. In the TWI method, steps are written in short sentences, followed by key points and reasons. For each of the tasks above, we have developed written TWI sheets that detail steps, key points, and reasons.

In the first few years, because we did almost all of our farmwork ourselves, I wasn't worried about codifying systems and passing on my practices. Eventually, we started adding interns and occasional paid staff. Usually, unless the task was very simple, I had helpers work alongside me, with training becoming observation for them and many little commands from me. The system worked, but it didn't make the best use of their time and skills or mine.

Now I use TWI sheets to remove the waste. First-quality work still doesn't always happen every time on our farm, but with better training we've increased work quality without adding work for ourselves. TWI helped me figure out the right kind and the right amount of information to pass along. In practice, I pull out TWI sheets when I orient new workers so that I don't forget to teach a step. If they've forgotten something, our workers can revisit the sheets, which we keep in a binder in our processing room. We are starting to rely on standard operating procedures (see below) to give simple harvesting instruction and to ensure everyone remembers standards after the training is done.

Our internship program is a core element of our farm. Interns have contributed a lot, but they come and go, creating a perpetual need to train new workers quickly. TWI makes this work easier. Here's the procedure we use when training with TWI sheets:

1. Explain all steps, key points, and reasons for each task, ideally while performing the task (this also demonstrates pace).
2. Work alongside new workers completing the task.
3. Allow workers time (ten to twenty minutes) to become familiar with the task on their own.
4. Observe workers, noting positives and areas for improvement.
5. Work alongside if necessary until workers can handle the task independently.

Our demeanor is important. If we are relaxed and encouraging, that rubs off in the form of motivation to perform a job well.

TWI Sheets

The fastest way to teach a task is to make sure training includes a breakdown of important steps plus key points and reasons for each step. Below is a sample TWI sheet from our farm—we fill out one of these every time we develop a new task. We also have TWI sheets for greens washing, greens harvesting, spray station, and microgreens/shoots.

TWI: GREENS BAGGING

Important Steps	Key Points	Reasons
A logical segment of the operation, when something happens to advance the work.	*Anything in a step that might* 1. *Make or break the job.* 2. *Injure the worker.* 3. *Make the work easier to do, i.e., "knack," "trick," special timing, bit of special information.*	*Rationale behind the key points.*
1. *Fill special orders first, then CSA orders. Rest goes to farmers' market.*	Use bags labeled 3#. Write "3#" and destination code. Scale target: 3# = 3.01–3.05	
2. *For farmers' market, put ⅔ of the mix into ½# bags, the remaining ⅓ into ¼# bags.*	Scale targets: ¼# = .26–.28 ½# = .51–.53 Place empty bags into tub to fill salad.	Much faster to fill a target range than a specific weight. Less movement to start with empty bags in the tub.
3. *Place filled bags into 2 black totes.*		Easy to slide multiple bags to other end of table for sealing.
4. *Seal with foot-pedal sealer.*	Seal line about 1″ from top. Keep as much air as possible in the bags. Depress pedal for one full second, let cool for one full second. While depressing pedal, grab next bag.	Air helps cushion greens. One second of depressing pedal and cooling creates wide, tight seal.
5. *Fill totes full with one weight class. Label contents on totes with sticky notes.*		Makes organization easier at market.
6. *Leave totes on table (winter) or move to cooler (summer).*	If above 50°F, take full totes immediately to cooler.	

2. STANDARD OPERATING PROCEDURES (SOPS)

Lean SOPs are instructional sheets that contain pictures— snapshots that show workers how you want a job done or a finished product to look—along with short texts explaining how to perform tasks in each picture. TWI sheets help trainers train; SOPs ensure quality. We use SOPs as a condensed version of TWI sheets.

According to Susanne Pejstrup, the lean farm consultant from Denmark, SOPs perform three functions:

1. They ensure all tasks are performed the same way by all workers.
2. They make it possible for farmers and workers to discuss and discover best practices. "It is not possible to make improvements if you are not sure of what you are doing now," she explained.
3. They make it easy to give instruction to new employees and for them to meet your expectations.

SOP sheets should be placed close to where a job is done. The text should be minimal—pictures should do most of the talking. To back up the SOP sheet, you can supply separate books and files explaining why. Since our farm is small and we are directly involved in most production steps, we don't rely on lots of documents to explain jobs. Still, having just a few SOPs and TWI sheets has made our management smoother. For larger farms with dozens of staff and training as a daily reality, the impact can be even greater.

We post SOPs in our processing area (see the photo of our task management board later in the chapter). They show our crops in the finished state, along with short texts explaining standards, such as how heavy a bunch of kale should be and how it should be banded.

3. VISUAL SYSTEM MANAGEMENT (VSM)

SOPs are examples of visual systems for running your farm. Pejstrup told me that in her work in Denmark she encourages farmers to use visual management as much as possible, because pictures are the clearest and fastest way to get your point across.

The Nordgaarden dairy farm in Næstved, Denmark, posts SOPs at eye level.
Photo by Susanne Pejstrup, Lean Farming®.

6a. Feeding - Correct removal with multi-purpose bucket

6.1a	**Make sure the bucket is clean** The grab must not be missing any tines, and bended tines must be replaced or straightened out	
6.2a	**Make a clean-cut and vertical silage face**	
6.3a	**Attention!** • Avoid uneven silage face • Avoid lifting up the silage and make sure to maintain a clean-cut silage face at removal	
6.4a	**Attention!** Avoid "biting off" silage with both top grab and bucket in the stack	
6.5a	**YES!** • Use "peeling" technique in maize silage • Place multigrab so the tines are in vertical line with stack and "peel" of silage • Keep the bucket out of the stack when removing silage • Afterwards collect loose silage with the bucket	

An example of an SOP sheet used in cattle operations in Denmark. Courtesy of Knowledge Centre for Agriculture, Denmark.

With oral management instructions are often forgotten or misinterpreted. With written management workers can waste a lot of time leafing through pamphlets and books.

We've been using more and more visual cues on our farm. I joke (only halfheartedly) that my goal is to run our farm for an entire harvest day with pictures instead of words.

More examples of VSM that work well on farms include:

Task Management Boards

A task management board tells you and your workers what needs to happen for a day, a week, or even several months in advance. It should be color-coded to indicate when tasks are completed.

I used to keep a weekly spreadsheet of projects, vegetables to harvest, and customer orders in my pocket. As part of our leaning, we decided to get the spreadsheet out of my pocket and post it in the form of a task management board so everyone on the farm could see it. Our goal was to be able to tell at a glance from a distance of 10 feet away how work for the day is progressing. To make that possible, we bought a used 3′ × 4′ magnetic dry-erase board along with magnetic grids and two-color indicator magnets. The whole project cost less than $100 and took just a few minutes to set up.

On the left side of the board, we list vegetables to be harvested or tasks to be completed. Across the top we list customer destination codes, each acronym representing a restaurant or farmers' market (FM) or CSA. My job as the order collector is to assemble as many orders as possible before the start of the workday. I then write the quantity down on the grid. For example, if we want to put turnips in CSA boxes that week, I'll write "70" (indicating the number of units needed) in the box that lines up with "turnips" and "CSA." I'll assign a person to harvest by writing his or her initials on the board next to an item.

The two-color magnets indicate progress. When the item is picked, a worker places the magnet in the box green side up. After the item is processed and in the cooler, the magnet is flipped over to show red—"stop"—telling me the item is market-ready. For tasks, I might write "Irrigate field for 30 minutes" or "Weed carrots in greenhouse B." Again, workers indicate completion with a red magnet.

The board is an effective way to manage the dozens of different harvesting tasks and small jobs that need to be done in the course of a day on our farm. As a manager, I like knowing what has happened and what still needs to be done. Workers like the clear instruction and being able to check progress. Also, if workers run out of things to do, they no longer have to chase me down to find out what's next—they just check the board. And it's fun. It turns work into a version of a board game: everyone is satisfied when the entire board is red.

The board streamlines communication. At the end of the week, I take a picture of the board with my iPhone and send it to three people. First, I forward it to the people staffing our farmers' market booth to tell them how much we are sending to market. At the end of market, they subtract what was left over and record total sales on the Google Drive spreadsheet for forecasting. Second, I forward the image to the person delivering our food to wholesale accounts as a checklist of what to deliver. Third, I forward the image to Rachel in the office, who uses the data to invoice customers and tell CSA customers, in a weekly newsletter, what they are receiving in their boxes.

We still sometimes use oral instruction or write out tasks or lists on slips of paper if that type of communication makes more sense. But usually the magnet board accomplishes our management job better.

T-Cards

Visual systems can get creative. Susanne Pejstrup described a system that a hog farmer uses based on a *kamishibai* board. *Kamishibai* is a form of storytelling by 12th-century Buddhist monks in Japan who used illustrations on paper scrolls

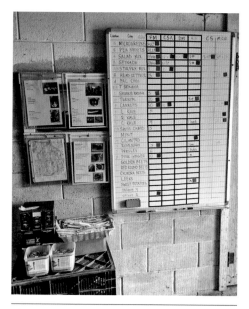

The lean task management board on our farm. A green magnet indicates an item was picked; red indicates it is washed and market-ready. SOP sheets hang on the left above packing supplies.

This lean board at Aluminum Trailer Company displays value stream data, 5S goals, types of waste, and training sheets.

to communicate to mostly illiterate audiences. The practice was revived in the 1920s as *kamishibai* storytellers traveled from town to town by bicycle. A version of *kamishibai* eventually became a tool in the Toyota production system: Toyota used cards placed in slots as a visual control for performing audits. The *kamishibai* board at the hog farm consists of dozens of pockets; t-shaped cards fit inside the pockets. The cards have tasks written across the top and printed on both sides, one side red and the other green. The cards are grouped under headings that show the time of day in which the tasks should be completed. The color facing out indicates progress (or lack of it).

Pejstrup explained that the farmer sees several benefits:

The *kamishibai* board at the Tummelsbjerg hog farm in Gråsten, Denmark. When tasks are completed, the cards are flipped. SOPs hang nearby. Photos by Susanne Pejstrup, Lean Farming®.

1. The board provides a complete overview: from a distance the farmer can see how far everyone has come.
2. The board saves time because the farmer doesn't have to check every corner of the property to see if employees have completed their jobs.
3. It is motivating for workers to turn the cards.
4. The farmer does not have to give as many oral commands throughout the day.
5. Employees can help each other complete tasks.
6. If not all tasks are completed, the team talks it over and either move uncompleted cards to the next day or agree for someone to put in overtime.
7. Employees can use the board as a checklist to remember their jobs for the day.

The farmer posts SOP sheets next to her board so that workers can quickly check how tasks should be done. Her board uses a fourteen-day rotation, but Pejstrup told me the system can work for any time frame. For example, you can use a task management board to block out large projects for an entire year. If in July you complete a task that was scheduled for August, you move the task to the July column and flip it over.

Visible-Target Management Boards

Also called metrics boards, visible-target management boards, like SOPs, should be posted at eye level. They help farmers track visible progress on their goals. Ideally, they are also relevant for workers. Just as workers can take part in creating metrics, they can help track the metrics. "It must be possible for the employee to make an impact on the figure and the curve," Pejstrup told me. This gives workers energy "and it is fun for the employees to reach goals together."

Produce farms might use visible-target management boards to track unsold goods and delivery errors (taking the wrong item or the wrong amount to the wrong place). Farmers and workers together can track the number of items that come home from the farmers' market unsold with a goal of seeing that line go down. They can count delivery errors on a weekly basis with a goal of seeing lower numbers over the course of the year.

Pushing Responsibility down the Ladder

If *kaizen* is a wheel, then staff is the axle fixed onto the wheel, causing it to spin.

Your people are often the best sources for new ideas because they are closest to the waste. In Japanese the word *gemba* is used simultaneously to

Supervising the Work, not the Workers

Jobs and work performance should be assessed on how much value they are adding, not on how busily a person appears to be working. Ohno explained his frustration with managers at a Japanese Toyota plant who were "deceived by the appearance of work":

> They say, "Look how hard our people work," but I tell them, "That is not called working. They just have fast hands.". . . Instead of trying something different, they think they will be more efficient if they make the young women work faster. They are busy supervising the motions, for instance, noticing that one young woman has slow hands, and do not have eyes to supervise the work.[3]

Farms do require workers who can hustle, and there is nothing wrong with making that clear to new recruits. But ultimately results matter more than an appearance of fast pace.

On our tiny scale, we often have only one or two workers helping us. Still, I like to manage by giving helpers a broad overview of what needs to be accomplished and then checking for quality as crops are washed or seedlings are transplanted, for instance. If work isn't completed in the time frame we need, the solution is rarely to make people work faster; rather, it's to redesign the job to make it less cumbersome.

refer to people on the shop floor and to the shop floor itself. According to lean thinking, the *gemba* is the place to look for improvements. Visual systems respect workers by keeping them informed. In return, workers help perform *kaizen*.

A few hours before this writing, I received my weekly market report email from Patricia Oakley, who was managing sales at our booth. She reported that sales were good but that we sent in too much salad and that one variety of kale had started to look limp by the end of market. She suggested that next week we decrease the amount of salad by nine pounds and that we bag instead of bunch the kale. How could I have seen these problems without her help?

Dave Van Belle at Van Belle Nursery has started instituting daily "huddles" so that "we can communicate up and down the company quickly." He also organizes monthly meetings, where top management shows the rest of the staff sales numbers. It's a chance for everyone to see what's happening across the whole company. For Van Belle, his people are his best resource, and he wants them to "all be aligned in the same direction. . . . We want everyone to feel included."[4]

The Laugh Factor

Humor creates a more relaxed atmosphere and opens staff up to share ideas. We try to make work fun, and we laugh a lot. Farming is full of enough drudgery—harvesting in below-freezing weather, working early while most people are still asleep, lifting heavy crates—so why not lighten the load with humor?

When I visited Randy Ewert at Bair Lane Farm, he laughed as he told me stories of wandering around his farm in search of tools or mistakes they've made over the years. "You have to be able to laugh at yourself," he said. "There should be a place for embracing the humanness of inefficiency." It was apparent that his humor and lighthearted spirit were key elements keeping his farm creative and thriving.

Dave Van Belle says that in their bigger operation, fun explains their success. "If you have fun and people are laughing,

Workers at Van Belle Nursery share a lighter moment. Dave Van Belle says that laughter helps attract workers and shoppers. Photo courtesy of Van Belle Nursery.

then we're going to have way better workforce morale and productivity." That Van Belle workers laugh every day is part of what attracts people to work and shop there. "This industry requires a lot of hard work and long days, so it's refreshing to people when they see our willingness to laugh at ourselves."[5]

This isn't to say that farms should be level organizations, with everyone trained to perform every task. Especially during harvest seasons, there is a place for temporary hires for specific jobs. But even temporary workers will be more motivated when they can see their role in creating something of value.

We make it clear to our workers from the start that we are interested in their ideas. Some of the best suggestions we've received have been from newbies who came to our farm with fresh eyes from fields as varied as mathematics and art. Not all suggestions are keepers. We are still the farm's gatekeepers, and we must decide which ideas to implement, but our workers are as close to our products as we are, so we listen to them.

One worker, after hours of pulling plants by hand, developed a system for removing harvested greens from the greenhouse by using our greens harvester to slice plants off at the ground and then prying them loose with a wheel hoe. Another worker developed a clothespin system for pinning

Workers have contributed to improving many systems on our farm.

restaurant orders within easy eyesight in our processing area. Yet another suggested our system of washing some root crops in the field rather than moving them to a separate spray area. Each of these ideas was the result of giving workers responsibility and engaging their minds, not just their hands.

Intern, Apprentice, Volunteer, or Staff?

We apply the same values analysis with workers that we do with our customers. That is, we discern what is most important to workers and try to give them what they value. The more precise we are, the better the experience for the worker and for us. We have noticed that workers in the different categories below come to us for different reasons. If we can be sensitive to what each type of worker is interested in, we can find tasks best suited to them.

Intern

An intern is exploring careers and interested in testing the waters in different settings. An intern might spend three months on a farm, three months at a doctor's office, and three months in an accountant's office. An intern is trading labor for experience. Their goal is to get a feel for the occupation, not necessarily to learn the specifics of the operation. For interns, I try to mix up their work so they experience different aspects of the farm, and I talk with them about the joys and challenges of farming as a career.

To attract the right interns, advertise the range of *experiences* you can offer them.

Apprentice

An apprentice has decided on farming as a career and is at your farm to learn how to space turnips, trellis tomatoes, and prune peppers in a greenhouse. An apprentice is trading labor for knowledge. With apprentices, we incorporate teaching and training as much as possible. When I am seeding a new crop, I might stop, call an apprentice over, and explain what I am doing. I'll go over our seed catalog order and explain why we use special techniques for certain crops. We

Clay Bottom Farm crew with Natural Resources Conservation Service representative (second from left).

also share with these workers more specific planning documents, such as planting charts, and we include apprentices in spring planting decisions.

To attract the right apprentices, advertise the kinds of *knowledge* you have to offer.

A potter friend of ours, Dick Lehman, visited several Japanese pottery studios midcareer for study and inspiration. The apprenticeship model has a long history in Japanese pottery, with apprentices sometimes working up to ten years for no pay. To keep apprentices engaged in spite of work that can be tedious, Japanese potters used the analogy of an empty cup. Your job is to fill an apprentice's cup with experience and information, noting whether a cup is too full or not full enough. Is it time to pour faster or to slow down?

Volunteer

A volunteer's goals are to have fun and make a meaningful contribution to a local farm. A volunteer is trading labor for a good time. Volunteers don't need in-depth explanations, just well-organized projects that allow them to get their hands dirty and feel as if they are making a difference. When we have day volunteers, I try to assign them projects like ripping out an old crop or transplanting a field of broccoli, so they can easily see the contribution they are making. As much as possible, I try to work with them or have them work in groups to make their time lively and enjoyable.

To attract the right volunteers, emphasize *fun* and the opportunity to contribute to a local food farm. We al-

ways try to include a snack or a meal to say thank-you and allow for relaxed social interaction.

Staff

A staff worker shows up in the morning and does what you tell the person to do. They are on your farm to trade labor for money. Although it is still important to sometimes teach staff beyond the training they need for a job, because you are paying them, their interest—and yours—is for their work to contribute to your farm's *profitability*.

To attract the right staff, offer compensation, hours, and benefits that are competitive for your area.

▦ ▦ ▦

We have used workers from each of these categories. When I meet with potential new workers, I try to discern their motivation for approaching us. Are they here primarily for an experience, knowledge, to get outdoors and have fun, or money? Can we offer them what they want, or might there be a better place for them to get what they are looking for? What type of motivation would best suit our farm at the time?

The right combination of intern, apprentice, volunteer, and staff will vary from farm to farm. There are costs and benefits with each category. Paid staff are obviously more expensive to retain, but the care and time required to treat unpaid workers well is a cost, too. Experiment with different types of workers to see what best suits your management style and the needs on your farm.

CHAPTER 10

Lean Applied at Clay Bottom Farm: Ten Specific Cases

"Follow the decisions that were made"...
It's easy to say these words, but not easy to do.

—TAIICHI OHNO

Throughout this book I've discussed lean principles applied on our farm and on other farms to show that lean is a system for everybody, from big grower to microfarmer. Below are ten case studies of applied lean thinking on our microfarm. While the specific solutions we've found for our problems will not universally apply to other farms—each farm is unique— the ways of thinking can. My aim in discussing these case studies is not to suggest that all farmers should grow their tomatoes or prep their fields like we do; we have not arrived at perfect and final solutions. Rather, the point is to demonstrate *habits* for finding more efficient ways to work, using lean vocabulary as an analytical tool.

Case Study 1: Leaning Up Our Land Management

My grandfather started out farming with horses, and my Amish neighbor, a dairy producer, still does. With a team of horses, your work slows down— plowing a field can take a week, several if rains interrupt the work.

The practice also puts a farmer in close contact with the soil, as horse-drawn plows tend to ride much lower to the ground than tractor seats. The proximity to the soil and the slower pace allow you to observe things well with your own eyes. So while horse plowing is slow, it is also lean, in the sense that farmers are noticing very closely what needs to happen and fine-tuning as they work. There is little wasted effort.

Once during a light afternoon drizzle I saw my neighbor Merle Miller and his team of horses plowing the field next to ours. To pull a single-bottom plow he used six horses, a considerable number. Miller told me that if their soils are lighter than his, farmers might use as few as three horses to pull a single-bottom plow. Even though the ground was wet, he kept going. At the end of each pass, he stopped at the edge of our property to let his horses rest. In hot weather he will let his horses rest at both ends of the field. In cooler weather they are allowed several passes before a break. "The key is to pay attention to your horses," Miller told me, "and not push them too hard." As Miller plowed, his eyes were on the ground. In sandy spots he adjusted his blade to plow deeper; in the heaviest spots he raised the blade. This helped keep his horses and plow in a straight line. Miller told me later that if you look closely, you can see his plow lines are not always perfectly straight. There is usually a little curving around the pockets of sandier soil and heavier soil, but by adjusting his plow he keeps the curves small.

I was also on my tractor that day, but not for long. Tractors have much less traction in clay soil than horses. Even a small amount of moisture can cause tires to spin. As mine spun, I tried going faster to keep momentum, but I knew enough about tractors and clay to know I was wasting time. I was using a 1953 Ford 8N that we had purchased for $3,000 at a local farm auction, along with an even older two-bottom plow that I'd bought from a local Mennonite pastor who didn't have time to farm. After just a few passes, I gave up the spinning, parked the tractor, and fiddled away the rest of the day in the shop.

Miller and I were both in our fields that day. I had more modern technology than Miller. But I had little to show for it. Next to Miller's, my field looked like a demolition derby ring. "With horses, it almost doesn't matter how wet your ground is," Miller told me. "I could plow through water from here to the hay barn."

So one of my first lessons in leaning up fieldwork was to learn how to slow down and closely observe, to adjust the equipment and approach to the ground condition. On a day with good conditions, my plowing might be faster than Miller's—Miller said it would take two to three days to plow 10 acres with a two-bottom plow and more than three days with a single-bottom plow. A tractor can complete the task in a few hours or less—but it will never be as precise.

Shortly after we'd moved to our current farm, Miller explained to me several lessons he'd learned about working the clay soil that is on both of our farms. He showed me how to watch for just the right amount of dried tips on clods of soil, which would indicate the clods would bust up well, making for a good disking day. "If you wait too long," he told me, "the clods turn to rocks. If you don't wait long enough, the soil just gums up the disk."

On the sandy land that I'd grown up on, the soil was rarely too wet to work. On our new clay, the workability window was tiny.

That first season, I plowed in the previous owners' soybean stubble and determined low areas that Rachel and I fenced off for pasture. We installed underground drain tiles in our vegetable production areas and underneath the building site of our future row of greenhouses. Our land management was traditional:

1. Plow to flip the soil and deprive the plant matter on it of sunlight and oxygen, in the late fall or early spring.
2. After a few days, once all green matter has died, disk to break clods into pieces.

We initially built up our soils by plowing in cover crops.

3. Pass over the disked field with a cultipacker, a heavy implement with steel rollers, to further bust up clods and level the ground.

For large-seeded items like corn or green beans, this amount of working the ground was enough. For finer seeds liked lettuce and turnips, we used a walk-behind tiller to pulverize the soil and create an even smoother seedbed.

To manage fertility, we relied on cover crops, rotating clovers, sudangrass, oats, vetch, rye, and field peas between our cash crops. Miller told me that his best corn always seemed to grow after his clover, probably because of the clover's ability to fix soil-bound nitrogen and make it available to crops. He never grows corn in the same field more than two years in a row before rotating to hay crops, which consist of clovers, alfalfas, and sometimes oats. When he has time, he plants oats in the fall after his corn harvest to add even more nitrogen. His crops and his soil are healthier for the practice. We saw the benefits of cover crops, too: soil that crumbled better and sturdier, more vigorous plants. Cover crops loosened up plots that were compacted and turned depleted fields into nutrient-rich soils.

HANGING UP THE PLOW

That traditional land management system worked well, and we saw steady improvement in our soils. But I began to wonder if a different system might be better suited to our small scale and the types of crops we were growing. On Miller's dairy farm it made sense to plow deeply and often: his crops—like field corn and alfalfa—were deep-rooted crops. Skillful plowing, along with disking and leveling, was a must if he wanted to plant a new crop the next year.

But on our microfarm, the root systems of our annual vegetables were not deep like corn or alfalfa. Most of our crops were being rotated in and out multiple times per season, sometimes in as little as three or four weeks. It didn't make sense to plow between every crop. It was too much effort. It involved a lot of passes over the field.

As I've mentioned elsewhere, a farmer's work in many ways is like setting a stage. In theater, stage managers lay out furniture and props, set up lights, and clean to get a set ready for actors to take over and create a show. Farmers plow, fertilize, set up irrigation systems and fences, and otherwise prep the stage of their farms for the real actors—the sun and the life within seeds and animals—to create the show.

I wondered, "Were there more efficient ways to set our farm's stage?" Plow farming, whether plowing in cover crops or corn stubble, involves at least two field passes—to plow and to disk—and possibly more, to create a smooth seedbed. I wasn't convinced that in our small operation we really needed to be spending so much time preparing our ground. What if we could cut our field passes—the number of times we passed over a piece of ground with an implement?

So I began experimenting with less invasive ways of incorporating crop residue and preparing beds for the next crop. Some crops I just disked without plowing. Others I shallow-tilled with a walk-behind tractor. Sometimes I covered crops with large sheets of plastic to smother them. By using these techniques, we were able to leave our plow parked for most of the spring and fall, except to incorporate the more deeply rooted cover crops. We had succeeded in cutting plow time in half.

But could we go further?

What if we could find alternatives to cover crops? Organic matter, which cover crops provide, is the blood coursing through the veins of our farm. Without it, crops on an organic farm don't thrive. But were there less laborious ways to supply our vegetables with decayed plant matter? Could we eliminate the plow altogether?

We began experimenting with deep applications of organic matter—not from cover crops but from compost that we made on our farm using grass clippings and leaf matter from our own pastures and from a nearby composting facility, as well as animal manures from neighboring farms. We started out with an inch or two of compost on a few beds, and we loved the

Compost making in action.

Mechanizing the 30-Inch Bed System

Like many produce farmers, we grow on beds rather than in rows. Except for a few wider beds in the greenhouse, all of our beds are 30 inches wide. Our aisles are 12 inches wide. For most people, the 30-inch width is scaled right because it is easy to straddle and even short arms easily reach the middle. Beds are also more space-efficient than rows, having fewer pathways, thus allowing small farmers to maximize their land and to concentrate compost. Larger farms often find row-cropping more efficient because long, straight rows are ideal for tractor-mounted tools needed for larger plantings and harvests.

In our first few years, we tilled with a hand-operated BCS tiller after plowing and disking, and we laid out our 30-inch

We now farm on heavily composted raised beds.

results: the compost added plenty of fertility and also tilth—the word farmers use to describe workable, friable soil. Compost busted up the clay.

There were other gains as well. We significantly reduced our cycle time, the total time required to produce each crop. Green beans no longer had to wait half a year for a cover crop to enrich the soil. And we found that by eliminating cover crops we'd nearly doubled the size of our farm because we didn't have to devote half of the farm to non-cash crops.

We were hooked.

beds with string. We raised them by shoveling soil from the aisles onto the beds (for root crops) or simply by walking the aisles. We spread compost by hand. If beds needed to be loosened, we used a handheld broadfork. The process worked fine, but it wore us out. Every step was accomplished through backbreaking handwork and involved a lot of lifting—a chore with our heavy clay soil. So we searched for a way to mechanize the practice. We wanted the same tight spacings but with less work.

First we invested in a compact tractor. We chose a Kubota L3400, with skinny 11-inch ag tires, because its small size would fit into our greenhouses and because the distance between the rear tires was 31 inches. We hung up our broadforks and started using a small, four-shank chisel plow pulled by the tractor. The plow is narrow enough to loosen just one or two beds at a time. We even use the chisel plow in the greenhouse, though we still loosen up corners and tight areas by hand. We purchased a 5-foot power takeoff (PTO) tiller for the tractor to till up larger areas. Whenever practical, we drive the tractor fast, in second or third gear, and slow down PTO speed in order to mix the soil without overpulverizing it. We use the tractor-mounted tiller for

most tilling but keep the BCS and the Troy-Bilt Jr. for tilling up small sections.

Instead of creating beds with shovels, we now use a bed shaper. We searched around for a compact model that would squeeze into small spaces and settled on the RB448 from Nolt's Produce Supply. The shaping pan is lined with ¼-inch-thick, heavy-duty, Teflon-like plastic so soil doesn't stick. The beds taper just slightly from the middle so that water drains off in heavy rains. And if we tilt the shaper an inch or two toward the tractor, we can actually create 6-inch-tall beds. If we want to lay plastic mulch, the shaper can lay it at the same time, along with irrigation drip tape.

In the tight quarters of the greenhouse, we still shape by hand or with a furrow attachment behind our BCS tiller, furrowing the aisles and raking two mounds together into a bed.

Through most of the year, we prepare an average of four 60-foot-long beds per week. A hand-prepped bed can take up to forty-five minutes to prepare; with machines the task takes fifteen minutes and requires much less human effort. This system works well for us in heavy clay. Other soil types might not require so much mechanization.

But I still wasn't satisfied. When we counted our moves, both cover cropping and composting required a similar number of steps, even though composting allowed us to permanently park our plow. Were there ways to make our composting more efficient and save effort when spreading it? We learned early on that shovels would not do. Our backs would not last long moving shovelfuls of compost at a time. We bought a small used skid loader from a neighbor, and to begin we trucked compost to the edge of our plots and spread it by hand to the middle.

Eventually, we traded the small skid loader for one that could move three times as much compost in the same number of trips. It was also wider and taller, so we could straddle our growing beds, keeping tire tracks in the aisles, and dump compost precisely where we needed it. This saved us from having to shovel at all except for a little leveling.

To reduce moves further, we decided to try applying 8 or 10 inches of compost, not just one or two. We started with just a few beds. The idea was to see if we could cycle multiple crops through the beds before needing to reapply compost. Remember that planting seeds and harvesting crops directly adds value for customers; playing around with tractors and skid loaders is type 1 *muda*—necessary, perhaps, but not part of adding value.

Our crops loved the deep compost. We'd never seen carrots as long and straight. Our greens were sturdy and warded off pests. Our radishes grew faster than we thought possible. *We* loved it, too. We were now able to cycle multiple crops with less work. After a few seasons our soil was so friable— we could easily stick our hand in to a depth of 8 to 10 inches—that we often were able to pull one crop and plant another without any soil prep at all.

THE COMPOST FARM

The deep composting system worked so well that we eventually covered our entire growing area, including in our greenhouses, with 8 to 10 inches of compost. Our farm now consists of fifty-six rich and loose semipermanent raised beds in the field, each 60 feet long, and fifty beds in the greenhouses, each approximately 40 feet long. We chose beds that are no longer than 60 feet mostly for psychological reasons—beds longer than that felt daunting to weed or harvest since we hand-harvest most crops, and the majority of our weeding is handwork, too. The beds are only semipermanent because we reshape them with a bed shaper (by hand in the greenhouse) once a year on average to pull soil that has eroded into the aisles back to the top of the beds. If we had sandy loam or another, more desirable soil, we might add less compost or rely more on supplements.

The system works well on a small scale. It might not work if we were trying to grow on 50 acres, where organic matter from cover crops might still pay off. We test our compost every few years, and we have become better at making custom mixes that work well for different crops. For vegetative crops like leafy greens, we use a mix with higher nitrogen from composted animal manures. For fruiting crops, the mix includes mostly composted green manures.

Compost is so central to our operation that we think of our farm as a compost-making farm almost as much as we do a vegetable farm.

Minimally Invasive Soil Management

As a general rule, we flip beds from one crop to another quickly, often on the same day. We pay close attention to the state of each bed on our farm every time we enter this transition. Does the bed need to be loosened with a broadfork, chisel-plowed, or tilled, or can the crop just be pulled out and another planted right away? We do as little as we need to.

For example, on one bed last year, I grew three successive crops of head lettuce. To harvest head lettuce we cut just below the soil surface—the roots stay in the ground; the heads are wiped with a rag and go into a bag. In this particular bed we transplanted immediately after each harvest, without any soil prep in between. The old roots just stayed in the ground to decompose and turn into food for microorganisms. Practices like these save time and also keep our soils biologically active and alive rather than overworked and tired.

When there are more residues to manage, we kill the old crop with plastic tarps or by tilling, and then we re-seed or trans-plant as soon as the residue is decomposed, usually in a few weeks. We will loosen the ground with a chisel plow or broadfork only if we are planting root crops or if the ground is very compacted. When we are in a hurry to replace a bed that contains a lot of residues, we use our root-digging implement or a wheel hoe to shallowly slice below an old crop, and then we remove the residues with rakes or by hand. If the crops are tall, we might mow them first with a lawn mower or string trimmer so they are more manageable to handle.

Between crops we also assess whether a bed needs more compost. We do this mostly by feel. Is the soil loose and friable enough for crops to grow uninhibited? We also look at the health of the previous crop. Were leaf tips turning yellow? Did the fruit fully mature? Over the years we've learned to trust our experience. Our eyes often can tell us as much as costlier soil tests. If a bed needs to be replenished with compost, usually 1 or 2 inches per year is enough to keep crops growing well.

Case Study 2: Cycle-Time Reduction in the Propagation Greenhouse

Our germinating greenhouse is the most valuable space on our farm. It contains expensive heaters, fans, and louvers to keep temperatures ideal for plant growth. We want to keep crops cycling in and out as quickly as possible. When we first started using the greenhouse, we spread out all of our trays horizontally on tables. Then we wondered if there might not be a way to grow up instead of out. We were heating the whole greenhouse, after all, not just the bottom half.

FASTER PROPAGATION WITH A GERMINATION CHAMBER

We started by designing a germination chamber that would allow us to get twenty trays started in the space normally needed for one—by stacking them up. A side benefit, we found out later, was that because the chamber provides perfect heat and air moisture, seeds germinated much better than before, and much faster. Seeds that used to take a week or more to germinate in less-controlled environments were now coming up in just a few days.

The first chamber we built and still use is an insulated bakery cart heated by a small pan of hot water. When the air temperature cools below a set point, water heats up, releasing both heat and

A farm-built germination chamber.

humidity. The environment is like the inside of a small sauna. We liked the chamber so well we built another one out of a used chest freezer turned on end. The new chamber can hold twice as many trays.

Large-scale operators use commercially built germination chambers, which are available through nursery catalogs and are a common piece of equipment in university biology labs. They're very expensive, starting in the thousands of dollars. Ours cost us less than $200 to build using parts from the local hardware store.

STACKED TABLES AND HANGING TRAYS

Another way we use vertical space in the greenhouse is to stack our tables once space gets tight. Crops requiring less sunlight can go on the bottom, while others go on top. We stack our tables two high so that the bottom surface is 28 inches off the ground and the top surface is 48 inches off the ground. We support them with cinder blocks, so the system is very modular and easily expandable.

The lower shelves provide a place to set trays in an area with diffused light. This works well for newly potted up peppers (for a day or so, while they recover from transplant shock), pea shoots, some microgreens, and other plants that appreciate a touch of shade.

We also hang metal pipes running lengthwise with the greenhouse to support trays over crops in the ground. For example, about 5 feet above newly transplanted tomatoes, we often set trays of early onions. The pipes hang from the greenhouse rafters. As long as the trays are high enough, the tomatoes still receive plenty of light.

A MINIGREENHOUSE FOR FASTER GROWTH

In the middle of winter, when we need to start only a few trays of tomatoes and peppers, there is no reason to heat the whole propagation greenhouse. So we built a greenhouse within the greenhouse. The minigreenhouse measures 6′ × 14′ and is constructed of 2′ × 4′ lumber. Because of a 30,000 Btu natural gas heater close to ground level, heat radiates upwards. A small fan circulates inside air.

We don't actually walk inside the structure; instead, we open large access windows to avoid heating aisle space. Rafters allow for easy hanging of grow lamps for the darkest weeks out of the year. The 7-foot-high roof is a flat piece of polycarbonate framed with lightweight angle aluminum that opens up when temperatures top 80 degrees F (27 degrees C). We use a shutter motor kit wired to a thermostat, placed at plant level, to control the roof vent. For extra bottom heat under young starts or heat-loving crops, we use a thermostatically controlled electric germinating mat under the seed trays.

Stacked tables and our minigreenhouse.

We experimented to find the right size for the minigreenhouse. If greenhouse spaces are too small, they heat up very rapidly and are difficult to ventilate. They also become too humid, which can lead to damping-off of plants. The square footage of our current minigreenhouse strikes a good balance: it's small and saves us money on heat, but it's big enough to prevent rapid heating and overly humid conditions.

The minigreenhouse does a great job with the early tomatoes and peppers, which we start in January. To fill the space while the tomatoes and peppers are growing, we also pack in early basil, microgreens, and shoots. Sales from these crops help offset a major portion of the cost of the heat. Also, on the north end, we built a multilayered shelf where we stack trays of pink ginger and turmeric, which need several weeks of 70- to 75-degree F (21- to 24-degree C) temperatures in order to sprout.

The minigreenhouse is an example of space-efficient, low-volume production.

Case Study 3: Minimizing Moves with Peppers and Tomatoes

Tomatoes and peppers are consistent top sellers for us, but the costs to grow them can be very high. Unless we take a lean approach, they lose money quickly.

We learned that customers value earliness on these crops and are willing to pay almost twice as much in May as in September. To increase earliness we moved production of these crops indoors, where we can better control the climate. We studied indoor production methods and got tips from nearby hydroponic growers (we still grow in soil) who had many more years' experience with greenhouse growing than we had. For the past six years, we've grown indeterminant tomatoes, which grow tall, and all of our peppers up strings that attach to wires running the length of the greenhouse. We prune every week to increase size and quality. The tomato strings are wound around spools so we can lean and lower several times throughout the year: by the end of the year, a plant can be 20 feet or longer, growing at a diagonal. We also grow determinant varieties, which reach a defined height, using a Florida weave trellising system. Since tomatoes are the most widely grown vegetable in the United States, there are many good resources on these systems. We did much research that paid off. Sure enough, better technique reduces work.

Still, there was a lot of waste in our production, which mostly showed up at harvest time in overproduction, overprocessing, too much motion, and overburdening (*muri*). Here's how we tackled these problems:

1. *Overproduction* waste showed up every August and September, when tomatoes and peppers seem to fall out of the sky in Indiana. While we have never grown the perfect, right amount, every year we get closer. With our standard red and yellow tomatoes, we put our focus on early and late sales through successive plantings of determinant varieties, whose production curve tends to looks like a crescendo: slow at the beginning, a steady climb, a peak, and then a decline. We plant four or five times on carefully chosen dates, starting in January in our mini-greenhouse, in order to stagger the peaks and to bump production in high demand times. We choose determinant rather than indeterminant varieties for our red and yellow tomatoes because of this ability to better control yields at different times. Production control is more difficult for heirloom tomatoes, which are indeterminant, yielding steadily throughout the season. We plant them just once and limit overproduction by planting a precise number based on previous years' sales. When we have excess of any tomato type, we sell in bulk to a local retirement village cafeteria or through a regional produce auction.

   For peppers, we simply hold back our harvest during slow times. Unlike tomatoes, peppers offer a long harvest window and actually increase in value as they turn colors.

2. *Overprocessing* waste showed up in our washing system and in our packaging. To reduce packaging waste we started bagging peppers in plastic bulk bags rather than in costlier wax produce boxes. (We now use bags for most wholesale items.) Tomatoes bruise in bags, so we still use boxes. However, we ask our wholesale customers to return used tomato boxes to us each week when we deliver their new tomatoes. We instruct whoever is delivering to ask for used boxes. This simple step saves hundreds of dollars each year.

3. We remove *motion* waste every year by finding ways to increase yield per plant. Having fewer plants saves moves. One example is pollinating. We found that hand-pollinating with a wand increased our yield significantly because more blossoms set fruit. This one change alone saved thousands of steps because we could now plant—and weed and prune and trellis—fewer tomatoes and peppers and achieve the same yield. We later replaced the wands, except for very early in the year, with a hive of bees that we set in the middle of one of our greenhouses.

   We also removed motion waste in our harvesting. We normally grow eight to ten varieties of indeterminant heirloom tomatoes and sell them as a mix. Previously, we planted each variety in its own row, harvested them separately, and then mixed them up as we put them in boxes. Now we mix them up at planting time so each row now contains

Heirloom tomatoes are intentionally mixed up at planting time to save moves in the packing house.

all varieties and so that we can sort and assemble variety boxes as we pick if we need to.

Another way we saved motion waste was to convert the milk-processing room of our barn into a climate-controlled tomato room. We used to cram all tomato processing into the same room where we sorted everything else. The room has two doors, so just-picked tomatoes enter at one end and ready-to-go orders leave at the other end. Our truck backs up to this second door.

4. *Overburdening (muri)* took the form of moving too many heavy containers. Most pickers instinctively want to start harvesting at the front of each row and work their way to the back. The result is that you often find yourself at the end of a long row with a heavy container that you need to hoist back to the front. We instruct pickers to start at the far end and pick toward the front so that they are more likely to end up at the front with their heavy load.

We experimented several seasons with different types of harvest containers. We settled on 5-gallon buckets for hybrid tomatoes because the handle makes for ergonomic lifting and they can nestle into tight areas. For peppers, which are lighter by volume, we harvest into the same 14-gallon totes we will take to market. Wherever aisles are wide enough—in the middle of the greenhouse but usually not along the edges—we use carts to set containers on as we harvest. However, our tight layout does not always permit such luxuries.

Case Study 4: Keeping Our Operation Small to Keep It Profitable

Just last week I received a text from a restaurant asking for an eclectic mix of vegetables in low quantities:

3 pounds of spring mix

3 pounds of spinach

one bunch of leeks

four turnips

one bunch of kale

one bunch of parsley

5 pounds of sweet potatoes

2 pounds of arugula

one ginger root

"a little bit" of rosemary

We specialize in these kinds of orders. When I set out to harvest this chef's order, I will make a quick loop around our farm, keep her list on my phone, and pack most of her order as I harvest. (See the discussion of single-piece flow in Chapter 5.) After the harvest I will walk just a few steps to our walk-in cooler and will need to drive only a few miles to get her order into her hands. In reality, I'll combine harvesting and delivering her order with several other orders.

Often I receive requests for our food from restaurants outside our self-imposed 10-mile radius. In special circumstances—if we have excess and the customer is willing to drive to our farm for pickup—we will fill the order. But usually we say no. We also receive requests for us to produce and deliver items not in our normal growing rotation: heirloom melons, for example, or purple asparagus. Again, we usually say no unless we are willing to source the item from another farm, serving as a go-between. We've been tempted to scale up. More than one landowner has offered to sell or lease to us more land at a bargain price so that we can expand our size. We say no.

Instead, we stay small. We choose crops and customers carefully and focus our energy on supplying a smaller number of customers with a wider but specific range of crops. We keep our operation small—and profitable—by spending a lot of time deciding what *not* to do.

▣ ▣ ▣

This is not to critique larger farms or to say small is the only route to profit. The *right* size farm is an individual matter, and our food system needs all types. However, it is a myth that economies of scale always pay off.

We decided to stay small, first, because we were making enough money to support ourselves and saw no reason to burden ourselves with more work. We also were drawn to the vision of a tidy farm that was pleasant to work on, where every step counted and every tool was well cared for. If we were constantly building and buying new things, cleaning up construction messes and drumming up new accounts, we wouldn't have energy to *refine* our farm into the shape of our vision.

Only later did we discover that staying small actually gave us an edge in certain aspects of production. Of course, there are efficiencies with big. There are efficiencies when planting 100,000 onions versus 100. Workers can get into an uninterrupted rhythm. Highly engineered machines—profitable only on a large scale—can perform tasks at lightning speed. Large growers can acquire bulk discounts on supplies that are unavailable to small producers. But there are downsides, too. Crops on gigantic farms are spread out, and produce could travel several thousand miles over a course of several days before reaching a paying customer, resulting in high transport costs. There are many miles between field and customer. Unlike small and custom producers, large operators can't understand what each customer values, since there are too many customers to account for. The distance between *everything* is great—between crop and cooler, between butcher and retail store, between the farm and its community of eaters.

Smaller operations can take advantage. We can optimize short distances to harvest and deliver products quickly. All of our crops are just steps, not miles, away from our processing area, thus fewer steps are required to transport from field to wash area. All of our customers are down the road, not across the country, thus we save tremendous transportation costs. We can minimize overproduction, with fewer accounts and less to go wrong. Small farmers can pinpoint investments and make sure they are getting the most out of every dollar. And by staying small, farmers can closely keep track of production. For instance, if we need to check inventories in the field, we take a short walk. For all these reasons, when our farm is running smoothly, we can produce very efficiently.

Small, local farms have the production advantage of closeness. Crops can grow close to greenhouses and processing areas—and are never far away from customers.

Taiichi Ohno writes of mass production versus low-volume production:

> *It is generally thought that mass production is cheaper. Thanks to this, although it is nothing to be thankful about, perhaps there is another bit of common sense that says low-volume production must be more expensive. However, when we question whether mass production actually reduces cost, I have to say that I have been around and seen a lot, but there were very few examples where increased production volume actually reduced cost. In most cases, increased production volumes increased cost.[1]*

Ohno used this misconception to Toyota's advantage. Because of the "commonly accepted belief that mass production is cheaper," Toyota was able to rely on low-volume systems and produce cars cheaply while selling at a higher price. "This can be extremely profitable," wrote Ohno.[2]

Small farms might enjoy less of a cushion to absorb shocks—if crops fail, small farmers can take a real blow. We can't take advantage of scale in production or in price haggling like big producers. However, every time my phone dings with a text from a local restaurant—from a chef who chose us over a faraway farm—I am reminded that small gives us an edge, too.

Case Study 5: Leaner Weed Management

Growing food on a farm crowded by weeds is like trying to drive with the parking brake on. Weeds drag down work and can grind production to a halt. For this reason, just like every other farmer we know, we aim to keep our farm as weed-free as possible. Over the years we've greatly diminished our weed population.

When we analyzed our weed management through the lens of lean, we realized our system was full of motion waste (too much handwork), overburden (too much effort), and wrong-sized tools. The best way I know of to attack a weed problem where it counts—at the root cause—is to ask why five times.

In Chapter 5 I offered the example of tomato sprout weeds in our winter greenhouses. Here's another example:

Problem: There is chickweed in the overwintered field spinach.

1. Why? Because we did not cultivate enough in the late fall.
2. Why? Because the task was overwhelming.
3. Why? Because the weed population was too high.

4. Why? Because chickweed lay dormant until the weather became cold, so we didn't notice it.

5. Why? Because we planted during warmer weather before the chickweed germinated.

The root of the chickweed problem was that we'd planted the spinach at the wrong time. Our solution is to not plant cold-season crops so late into the year unless we are absolutely confident that a plot is chickweed-free. As this example shows, hoeing, while at times essential, is often just a short-term fix, not addressing underlying problems. Here are other strategies that we use to keep our weeding time to a minimum:

1. *We put into cultivation only an amount we can manage.* Weeds tend to take over plots that have been forgotten. Weeding is a great group project, and we often begin a workday by starting the entire crew out on the east end of our growing area and weed entirely to the west end. If we've planted more than a small crew can cultivate in a day, we're in trouble.

2. *We establish perimeter control.* Once a week we use our Troy-Bilt Jr. to shallow-till quickly around the perimeter of all of our plots so weeds don't sneak in. Because the tiller also cleans up our aisles, the only manual weeding left to do is in the actual beds.

3. *We smother weeds with compost.* When we apply compost, which we aim to keep weed-free, we leave it on the soil to smother potential weeds rather than till it in. Tilling has the effect of bringing buried weed seeds up where soils are warm enough for them to germinate. The less we deeply till the better.

4. *We use mulches on long-season crops.* Rather than contend with weeds, we've found that it is worth the effort to grow tomatoes, peppers, eggplants, leeks, and other long-season crops under plastic mulch or landscaping fabric. Our general rule is to mulch any crop that will be in the ground longer than two months.

5. *We change crops quickly.* If plots are always in production, there is no time for weeds to take over. This is one advantage to keeping beds full of crops almost year-round. For a few months in the winter, some outdoor beds do not contain crops—we cover those beds with black plastic tarps to kill weeds and pre-warm our soils for spring.

6. *We standardize weeding.* We keep several green totes around the property and hang a couple of our favorite hoes in the middle of our growing area, which makes weeding always convenient to do. Most weeks we start out with a routine Monday weeding, so weeding stays at the top of our to-do list. We like to cultivate before weeds are 1 inch tall, since it takes much

less effort to hoe a weed that has just germinated than to yank out a waist-high bush. Time management is our most powerful weeding weapon.

Case Study 6: How We Space Farther Apart to Increase Yield

The key to producing more and more with less and less is to increase yield per square foot.

While we happily use a six-row precision seeder for densely seeded greens, for crops like radishes, beets, turnips, carrots, peas, edamames, and green beans we now frequently plant seeds by hand or in the greenhouse as starts so that we can space with even greater precision. Surprisingly often, we often find that wider, not denser, spacing provides higher yields.

We use a rake to grid out a precise pattern for workers to follow. I keep a list of our spacings on my phone, always at my side. For example, for large-size greenhouse carrots I use our 30-inch bed rake with 6-inch PEX plumbing tubes as row markers to grid out a 7½″ × 7½″ pattern on a 30-inch bed top. This gives us five rows per bed. Then I tell workers to put a seed at every intersection and five seeds between each intersection that runs the length of the row. This spaces carrots 1½″ apart and, if the soil is prepped right, will produce uniform and pristine carrots that we'll have no trouble selling. We used to thin crops to achieve this spacing, but after I timed two workers thinning we stopped because it took three times as long to thin as it did to seed correctly

Hakurei turnips individually spaced.

by hand. Even a finely tuned seeder cannot achieve that level of precision.

The extra effort might seem daunting and might not pay off on larger farms, but we've seen such a dramatic jump in yield—sometimes doubling our harvest per square foot—as well as in quality that we now put in the extra time up front.

One way to really maximize space is to cluster-seed crops like beets, turnips, basil, and spinach into plugs, using four to five seeds per plug, and transplant them into an 6″ × 6″ or 9″ × 9″ grid rather than direct-seeding with a seeder. Transplanting expands the size of the growing area at no cost because while your plugs are growing you can use your beds to grow other crops.

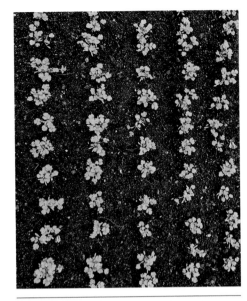

Cilantro transplanted by hand.

For leeks and onions, which we transplant into plastic mulch, we use a heavy steel wheel with removable prongs to poke perfectly spaced holes in our plastic—every 6, 12, or 24 inches depending on how many prongs we've left in. This is a scaled-down version of a waterwheel transplanter found on bigger farms. I made wooden handles for the tool that angle off to the side,

A steel wheel for spacing.

making it easy to push from the pathway. To transplant bigger crops like tomatoes and peppers we sometimes use a backpack transplanter that will poke holes in the ground or in plastic and release water into the hole. To plant single-seed crops like beans or edamame into plastic, we use a single-seed planter. One person jabs the tool into the ground while another puts a seed into the seed hopper.

Because we have so often seen yields jump when we get spacing right, we recently invested in a Japanese paper pot transplanter that can achieve precise transplant spacing—2, 4, or 6 inches—with less labor, and we plan to use it on as many crops as possible.

Table 10.1. Spacing of Selected Crops at Clay Bottom Farm

Crop	Transplant (TP) or Direct Seed (DS)	Rows per 30" Bed	In-Row Spacing
Basil	TP (4 seeds per cell)	2	6"
Beets	TP (4 seeds per cell)	3	6"
	DS	3	2.5"
Bok choy	TP	3	6"
Broccoli	TP	2	18"
Carrots	DS	5	1.5"
Eggplant	TP	1	18"
Fennel, bulb	TP	3	7.5"
Garlic	DS	4	7.5"
Ginger	DS	2	18"
Head lettuce (baby)	TP	3	6"
Head lettuce (full size)	TP	3	9"
Kale, summer	TP	2	9"
Kale, winter	TP	3	6"
Kohlrabi	TP	3	7.5"
Leeks	TP	4	6"
Parsley	TP	3	5"
Peppers*	TP	1	9"
Romaine	TP	3	4"
Shallots	TP	4	2.5"
Tokyo bekana	TP	3	6"
Tomatoes, heirloom*	TP	1	9"
Tomatoes, hybrid	TP	1	18"
Turnips	TP (4 seeds per cell)	3	6"
	DS	3	2.5"

* These crops are pruned to two leaders.

Case Study 7: *Kaizen* in the Microgreens Greenhouse

Microgreens, pea shoots, and sunflower shoots consistently scored high for us using our dollar value per square foot metric (see page 182), but our process for growing them wasn't always lean. There was a lot of defect waste: crops we couldn't harvest because of mold or seeds that didn't germinate. And too many moves: we filled 2-inch trays with expensive potting soil and moved trays or other items several times before harvesting.

After a few *kaizen* sessions, our defect rate is now less than half the previous rate, and we've cut our moves from twelve down to eight. In addition, we found ways to save costs on seeds, trays, and growing media.

We used to buy what sounded fun to grow, no matter the cost. Now we set a boundary: to make it into our rotation, seeds must cost $30 per pound or less. (Some seeds can cost well over $100 per pound.) After trying out dozens of varieties, we settled on robust but affordable field radishes and field peas (we include peas in our micros mix) that we buy in bulk from a local supplier, as well as a Chinese cabbage, a red radish, and red mizuna. We clean out all of our garden seeds every winter, and many leftover seeds from our field crops—like broccoli, arugula, and kale—make it into the microgreens mix as well. This is a great way to eliminate waste and use up old seeds. We continue to trial varieties, and I'm sure our mix will evolve.

Because microgreens and shoots are harvested young, they need very little soil. We started with 2-inch trays because that's what we had on hand. Our first move was to switch to 1⅛-inch trays, which we ordered through our local produce supplier. The trays are sturdier, last longer, and require almost half the amount of growing medium. We fill trays to the rim with medium, then sprinkle seeds right on top without covering them.

We used to grow these crops using our standard purchased growing medium, an organic blend from our supplier. But we decided one season to try different media that were cheaper and more locally available, such as garden soil and our own compost. We settled on our own compost because it was cheap, always available, and had plenty of nutrients for the young greens.

Having tackled defect waste, we set out to reduce moves. Our old sequence was as follows:

1. Move compost to the greenhouse.
2. Unload compost.
3–4. Load trays with two shovelfuls of medium (two moves per tray).
5. Move trays to benches.
6. Seed trays.
7. Move trays to our minigreenhouse for germinating.
8. Wrap trays with plastic (to hold moisture in).
9. Unwrap trays once seeds pop.
10. Move trays back to the benches.
11–12. Hand-water at least twice (usually more).

We found ways to combine steps and cut moves. First, with the shallower trays we use only one shovelful of medium per tray. Instead of moving heavy compost to the greenhouse, we now take our lightweight trays out to the compost heap and load trays straight onto the back of our John Deere Gator. As with our practice of loading CSA boxes (see Chapter 5), this means we are moving lightweight objects (trays) while keeping the heavyweight compost stationary. We now seed trays while they are on the

We switched to using shallow trays and our own compost to cut costs on microgreens. Photo by Emma Gerigscott/Clay Bottom Farm.

Gator. Then, instead of germinating trays in the minigreenhouse, we switched to using our new germination chambers, preheated to 78 degrees F (26 degrees C). The chambers are very humid, so there's no more need to cover trays with plastic. After thirty-six hours, we move trays to their final location. We set up an automatic watering system, making it unnecessary to hand-water except to touch up edges. Our new sequence is as follows:

1. Move trays to compost heap.
2. Fill trays.
3. Move trays to Gator.
4. Drive Gator to greenhouse door.
5. Seed trays on the Gator (unless it's raining).
6. Move trays to germination chambers.
7. Remove trays from chamber and move to benches.
8. Hand-water to touch up edges.

Case Study 8: *Heijunka* All Four Seasons

We leveled our load (the practice of *heijunka*) by discovering ways to extend fall crops later and start spring crops earlier every year. To start, we used simple structures like steel hoops with row covers and sometimes plastic to protect crops in freezing weather. We planted in the hoops crops like spinach and carrots in late fall and early winter to harvest in the spring. Also, we started spring transplants in midwinter so we could set plants out under the hoops early in the year. Thus, fall and spring met in midwinter, and our season became year-round. The field hoops were economical, but they were temporary, meaning they involved a lot of work to build and tear down every season. We eventually built our four permanent greenhouses, installing heaters in two of them, to replace the temporary structures, which we no longer use.

While we enjoyed winter growing, we quickly noticed that winter production was more costly. Crops took longer to grow, tying up space and increasing risk of defect. Freezing temperatures blocked up hoses with ice, so that they had to be moved or drained every time we watered. Our unheated greenhouses alone did not always provide enough warmth to keep a wide range of crops alive, meaning we needed to purchase agricultural row covers or pay for heat. Fingers and toes got cold quickly, slowing down work. Our processing room wasn't insulated well enough for workers to be comfortable. Ice-cold washing water forced us to install a water heater. And coolers required heaters to keep crops from freezing.

That's a long list of costs.

We were able to raise prices slightly to cover some of these added costs, but not all of them. To increase our winter profits, we had to get lean. For us, winter farming is still not as profitable as summer farming, but the benefits of a more level load make up for lower margins. Below are steps we took to cut waste from November to March:

By covering carrots with row cover and straw we can harvest after the ground freezes.

1. We *applied the "don't overdo" principle* to our winter crops. For example, in the spring, summer, and fall we rinse greens like mesclun mix and spinach. Competition for fresh food is high during those seasons, and customers choose clean products. However, washing takes a lot of time and from November to March would be uncomfortable, requiring hands in cold water. In the winter we started selling greens unwashed. This cut our post-harvest costs on these crops in half. Demand is high enough that customers are grateful for fresh food; they don't care that it isn't washed. We also limit our size options. Most of the year we sell greens in 3-pound, ½-pound, or ¼-pound bags. In the winter we sell only quarter-pound bags, which offers us a higher return.

2. We *cut greenhouse heating costs* by closing up air holes in the endwalls and pinning our sidewall curtains with wiggle wire, a special wire used to attach greenhouse covering to the frame.

   We keep our heaters set to 30 degrees F (-1 degree C), which is just enough to keep our cold-hardy crops alive. The natural gas heaters do require costly fuel, but the costs pay off in three ways: (1) crop growth is twice as fast with minimal heat versus no heat; (2) we save money on row covers (which we still use in the unheated greenhouses); and (3) we save time covering and uncovering row covers to harvest and ventilate. The heaters save us a lot of moves.

3. We *winterized our processing room* by blowing foam insulation on the walls and ceiling. A small gas heater set to low keeps pipes from freezing and keeps workers comfortable, which speeds up their work. A water heater means water temperatures are comfortable, which makes it much faster and more pleasant to hose down totes and clean up.

4. We *dedicated row covers* for the unheated greenhouses. These row covers are stored in the greenhouses and never used outside, which keeps them from wearing down in the weather.

5. Except on tender transplants, we *did away with wire hoops* in the greenhouses. In our experience, crops do fine with covers resting on top of them rather than suspended over them. There is occasional loss from covers freezing to leaves but not enough to justify the cost of hoops or the time it takes to set them up. We do still use wire hoops in the case of tender transplants that could break with the weight of a cover.

6. Because production costs and defect rates are already high in the winter, we've learned to *be conservative with experiments*. We're grateful for the efforts of universities and other programs that try out alternative heating systems, new crops, and experimental production techniques. We rely on their information. But on the production farm that must pay its own way, we've discovered that it's best to stick with tried-and-true varieties and techniques. Crops and seasons with low margins require a level-headed do-what-works approach or else those margins quickly evaporate—or freeze-dry, as the case may be.

   For us, this means growing a lot of spinach, which thrives in the cold; carrots, which keep well in the ground; and cold-hardy Asian greens. We grow all of these crops using conventional greenhouses, tested seeding schedules, and standard production techniques. We adjust the formula of our mesclun mix in the winter so that it includes more Asian greens and spinach and less lettuce, which is more difficult to grow in winter.

7. To save moves, we *automated ventilation and watering systems* in the greenhouses. When we counted our steps, we found that growing in

Timing for a Level Load

We've found that the trick to supplying crops year-round is timing. We are often out planting at very unconventional times. For example, we seed carrots in the greenhouse in mid-November for early summer harvest. We sow them outside in February (weather permitting) for midsummer harvest on raised beds that we've prepped in the fall.

For winter sales, July, August, and September are key planting months. In early July we plant kohlrabi, green onions, beets, and other field crops for a fall harvest. In early August we seed into trays Tokyo bekana, bok choy, tatsoi, head lettuce, kale, turnips, parsley, and Swiss chard, which we will then transplant to our winter greenhouses. In September we direct-seed in the field, and eventually in the greenhouse, spinach, arugula, mesclun mix, and other greens.

If we are off on our dates by even a week, we might miss our harvest targets. Each region is unique, and each farm a microclimate, so I suggest new farmers try

several dates the first year or two and keep notes. The table below is an example of how we tested seeding and harvest dates, in this case from 2010 to 2012, on baby lettuce and spinach.

Table 10.2. Clay Bottom Farm Winter Seeding Chart (2010–2012)

	Planted	Harvested	# of Weeks
Baby Lettuce	1 Sep	5 Oct	5.0
	19 Sep	23 Oct	5.0
	28 Sep	12 Nov	6.0
	4 Oct	12 Nov	5.5
	11 Oct	19 Nov	5.5
	20 Oct	19 Nov	5.0
	22 Oct	19 Nov	5.0
	24 Oct	9 Dec	8.0
	24 Oct	1 Feb	13.0
	16 Nov	3 Mar	17.0
Spinach	19 Sep	30 Oct	5.0
	27 Sep	10 Nov	6.0
	3 Oct	17 Nov	6.0
	4 Oct	6 Nov	4.0
	11 Oct	20 Nov	6.0
	20 Oct	26 Jan	13.0
	22 Oct	21 Jan	12.0
	24 Oct	15 Feb	15.0
	3 Nov	1 Mar	12.0
	11 Nov	1 Mar	15.0

greenhouses involved a lot of walking to open and close doors and curtains and perform watering tasks. We were constantly battling overheating by manually raising doors and curtains to let hot air escape.

When temperatures dipped, we rushed to close the same doors and curtains. Watering was also a chore. Soil can dry quickly in a greenhouse, and we were constantly hand-watering to make sure plants didn't die. These types of actions are type 1 *muda*—they lead up to value, but they don't add actual value; they are ripe for elimination.

We purchased automated louvers and endwall fans for all of our greenhouses. Many we picked up used at auctions; others we bought from greenhouse suppliers. They keep temperatures steady, on their own, for much of the year. We still use manual curtains and doors during summer for maximum ventilation. Thermostats on panels in the middle of each greenhouse control the fans and louvers, which keep temperatures below 80 degrees F (27 degrees C). Likewise, we installed overhead sprinklers and now use simple water timers. No more constantly wrangling with hoses and walking to the hydrant every time we need to turn water on and off.

■ ■ ■

A big bonus with winter sales is that we are maximizing fixed costs because we are putting to profitable use infrastructure—greenhouses, a processing room, a tractor—that we've already paid for and that otherwise would sit idle. We keep all four of our greenhouses full all winter with root crops and greens.

In practice, we've found that while crops do experience some winter growth, most actual growing takes place in the late fall. We then "store" these crops in greenhouses until we need them for market. In this sense our winter greenhouses are aboveground root cellars. That is, instead of harvesting in October and packing crops away for midwinter sales, we leave them in the ground, covered by a greenhouse, and harvest them fresh. Eliot Coleman and others have written extensively about winter growing—see Resources for Further Study for suggested reading.

Farming is by nature a seasonal activity because the sun is the engine powering a farmer's work. The sun is the essential ingredient awakening seeds and enabling plants to transform rays of energy into food for humans and animals. When the sun shines less, the farm *should* slow down.

But we humans have never been satisfied to rest when nature suggests we should take a break. As Thoreau once wrote, "I love the winter, with its imprisonment and its cold, for it compels the prisoner to try new fields and resources."[3]

For modern farmers, advances in technology—including many very low-tech innovations—and an ever-deepening understanding of plant and animal behavior in cold-weather climates have made it easier than ever to

Minimally heated greenhouses allow for efficient winter production and consistent quality.

push the boundaries of the traditional farming season. These advances offer hope that farmers in northern places might once again feed their communities year-round, while at the same time making their businesses leaner because their workload is more level and sane.

Case Study 9: Metrics to Pursue a Vision

All along, we wanted lean to help us organize our work in the pursuit of a vision: a small, well-kept farm where every step and square foot counted and where we are moving around lightweight but high-value items. We chose the metrics below to steer us in that direction.

DOLLAR VALUE PER CONTAINER

All of the totes we harvest into and take to market are the same 14-gallon size, with the exception of containers for tomatoes, which require softer cardboard boxes. This uniformity allows us to track the dollar value in each tote. Some crops (shallots and radishes) can fill a tote with more than $100 worth of product. Other crops (watermelon) barely fill a tote with $20 worth of product. That's a big difference when you are moving thousands of totes throughout the season. Since we want to be lifting lighter, higher-value loads, we use this metric as a simple way to calculate the per-volume value of our food. We focus our growing on crops at the top of the list to stay in line with our vision.

This metric is right for us—it fits our particular vision. It would not work for all farms. As I mentioned earlier, we partner with farmers who specialize in heavy crops that we have dropped from our list, and they are doing quite well.

But on our small scale, we prefer light loads. Remember in Chapter 1, when I discussed our relief after removing wagonloads of junk off our property? We experience the same feeling every time we replace a heavyweight item with a lightweight item. Our farming is literally lighter and easier to do.

YIELD PER SQUARE FOOT

We wanted to know the dollar value per square foot for each of our crops in order to have a way to know whether we were maximizing the fixed costs of our land, especially our greenhouses. To find this figure we simply harvest a section, add up the dollar value, and divide by the harvested area. For tomatoes or peppers, it is easier to track total sales for the season and then divide by the square footage the crop used in the greenhouse.

Table 10.3. Approximate Dollar Value per 14-Gallon Tote: Selected Crops at Clay Bottom Farm

Ginger	$360
Garlic	325
Shallots	225
Green onion	150
Kohlrabi	138
Radishes, bunched	125
Peppers, colored	120
Carrots, bagged	100
Sugar snap peas	90
Salad mix	80
Spinach	80
Romaine	75
Rhubarb	75
Potatoes, new	75
Microgreens	70
Pea shoots	70
Spring mint tips	70
Red beets, bunched	63
Fennel	63
Turnips, bunched	63
Carrots, bunched	60
Kale/Chard	60
Peppers, green	60
Potatoes, storage	40
Squash, butternut	40
Head lettuce	38
Bok choy	38
Onions	30
Watermelon	20

Note that yield per square foot does not equal profitability. For example, some items like ginger will eat up greenhouse space for two or more seasons. In order to earn their keep, they must have a product value at least double that of a single-season crop. In reality, value stream maps, which take longer to calculate, are the best way to tell exactly what your crops net per hour of labor.

For us, though, yield per square foot data is extremely useful on its own to help us decide which crops do not fit on our small-scale farm. Our fifty-six field beds and fifty greenhouse beds are precious territory. From each bed we aim to grow three to four crops per year, sometimes more in the greenhouses. With some of these crops, like salad greens and kale, we achieve multiple harvests from one seeding. Using the yield per square foot metric, we know which crops give us the highest return per bed. We aim to grow crops that return at least $2 per square foot, or $300 per field bed. If a crop cannot produce solid returns per square foot—cannot pull its weight—we don't grow it. Such was the fate, for us, of storage potatoes, butternut squash, and watermelon, among others.

Again, we partner with other farms that can grow these crops more efficiently than we can in order to supply the items to our customers. For these farms, with more labor, land, and bigger equipment, yield per square foot is less relevant. Also, while we normally stick to crops with high return per square foot, there are a few times in a year when we might make an exception. For instance, since we see broccoli as an essential CSA item to

Tracking dollar value per square foot helps us maximize a small plot of land.

keep boxes interesting, we grow just enough for the CSA, although not for our other outlets.

This kind of harsh math helps us make a living on less than an acre. In fact, I think we are just scratching the surface of what's possible. An acre is roughly 43,000 square feet. Walkways and access lanes—even with intensive cropping systems—will consume about one-third of that, leaving around 30,000 square feet for growing crops. What if we were truly utilizing all 30,000 square feet? Two harvests per year of an item that yields $2 per square foot, on the low end of our scale, should theoretically yield a farmer $120,000 per year per acre.

Unfortunately, farming with pencil and paper is much simpler than farming with shovel and hoe. In reality, crop losses, mistakes, erratic weather, left-over crops, or any number of other factors stand to foil neat plans. Still, the application of a little math will go a long way toward discovering the most profit-able uses of a small plot of land or a chicken house or a dairy barn or what-ever fixed cost you want to maximize.

FIELD TO COOLER IN ONE HOUR

Table 10.4. Approximate Dollar Value per Square Foot: Selected Crops at Clay Bottom Farm

Heirloom tomato	$19.25
Hybrid tomato	12.00
Ginger	12.00
Pea shoots	10.00
Salad mix	10.00
Spinach	10.00
Spring mint tips	7.50
Romaine	5.00
Carrots, bunched	4.50
Carrots, bagged	4.50
Shallots	4.50
Microgreens	3.75
Rhubarb	3.75
Turnips, bunched	3.30
Garlic	3.00
Red beets, bunched	2.80
Fennel	2.80
Kohlrabi	2.80
Head lettuce	2.50
Green onion	2.50
Bok choy	2.50
Potatoes, new	1.30
Broccoli	1.25
Sugar snap peas	1.25
Onions	1.00

For each of our crops, we like to know the field-to-cooler value, that is, the dollar value one worker can harvest, wash, and place in our walk-in cooler in one hour. Since we spend most of our time harvesting, this information helps us eliminate crops that are overly slow to harvest given their return and shapes equipment-purchasing decisions.

Field-to-cooler information also constantly reminds us that one of the best things we can do to increase farm efficiency is to improve our growing technique so that workers have a healthy stand of crops from which to harvest. Wormy kale can take more than twice as long to sort and pack as healthy kale.

Keeping a Kitchen Garden, Too

We keep a small kitchen garden on our farm, growing small quantities of items just for us, as a way to keep the soul in our homestead. The kitchen garden is exempt from the metrics that we use in our production space. We do with it what we please.

A kitchen garden improves our quali-ty of life. We are more relaxed about and creative with our household food pro-duction if it is in a separate space. We can get away from the hustle of running a business, and we reconnect with why we started farming in the first place: because we love to grow, cook, and eat our own food.

The field-to-cooler metric needs to include *everything* required for harvest and packing, beginning to end. For example, with tomatoes, the clock starts with gathering buckets and gloves for harvest and includes the time assembling cardboard boxes for delivery, sorting, weighing, and labeling. The use of this metric has shown us how important it is to store tools close to their use and has encouraged us to create smooth flow in harvesting.

Because there is not enough time to grow all possible crops all year, we use field-to-cooler information to let us see when to prioritize production of certain crops. For instance, while salad mix and spinach are profitable in the colder months, we scale back our summer production, when our time is better spent focusing on more profitable crops like tomatoes and peppers.

Case Study 10: A *Kanban* System to Replace Empty Beds

Kanban signals, visual cues that say it's time to replace, helped us organize our seeds so we are ordering them in timely fashion (see Chapter 5). Could we use the concept in the field to help us plan our growing?

One advantage of our standardized bed system is that we can harvest one bed, or part of a bed, and replace the harvested crop with another without having to rework a large area. We aim to keep all greenhouse beds full all the time, and we aim to keep outdoor beds full except in midwinter.

We don't spend a lot of time planning what will go where. We know from previous years how many beds of each crop we are likely to need, and we do not bother with complex crop rotation; because we grow such a wide variety and move crops around so frequently we've never seen problems due to lack of rotation. We simply start the year with a list of crops and the approximate number of beds per season that we project we'll need. We are a bit more careful with plotting out the greenhouse space. We want to place tomatoes, peppers, turmeric, and ginger in heated greenhouses and rotate them. Other than those crops, all we have to know for greenhouse crops, like field crops, is the number of beds, not their precise location.

As mentioned, throughout a growing season, as plots open up, we try to replant on the same day that we harvest and remove a previous crop. Rarely does a bed sit open for more than a week, except for very early or very late in the year when we run out of crops that meet our profitability metrics.

The open plots become a visual queue to replace—a *kanban*. Crops in the field, then, are like parts in Brenneman's bins at the trailer factory: when a parts bin empties out, we replace it. The difference is we don't always replace harvested items with the same item. Rather, we plant whatever is most profitable to put in at the time. The question we ask ourselves whenever a plot opens up is, What would our customers most value in this space at this moment? The answer changes all the time because the market is always changing and competitors are always up to new tricks. As a general guide, we do keep handy a very rough seeding schedule for the field and greenhouse, as well as our list of how many beds of each crop we forecast we will need in a year. But when planting time arrives, we balance those plans with an assessment of current reality.

This *kanban* style of farming has been key to our high profit margins because it forces us to use several lean tools for smooth flow:

1. *Kanban* planting guarantees we are maximizing fixed costs. All plots are in use.
2. *Kanban* planting gives us an opportunity every time we plant to replace lower-profit items with higher-profit items. The system is very flexible. Rather than impose a rigid schedule and field map, we use *kanban* to quickly respond to conditions on our farm and in the marketplace.
3. *Kanban* planting saves a significant amount of time designing our work plan for the year. Planning work is type 1 *muda*—necessary but not adding value. The *kanban* approach helps us minimize it.

Kanban bins at Aluminum Trailer Company.

Open beds become a *kanban* signal—we replant them as quickly as possible.

My goal in sharing these case studies is not to suggest that these specific strategies will work on every farm. While we strive to stay current and use modern equipment and efficient practices, there are surely better techniques we have yet to discover. Instead, my goal is to show a method of decision making that has worked on our farm. Rather than grow our business through hasty expansion, massive scale, and high volume, we grow through leaning. Our success is the result of low defect rates, efficient process, tight production control (minimal overproduction), low inventory-carrying costs, high fixed-cost utilization, a level production and sales load, and a commitment to rapidly changing what we do. Our basic practice is to keep our customers and what they value always in our minds, then to remove impediments to achieving that value from the minute we order seeds to the moment we hand customers our food.

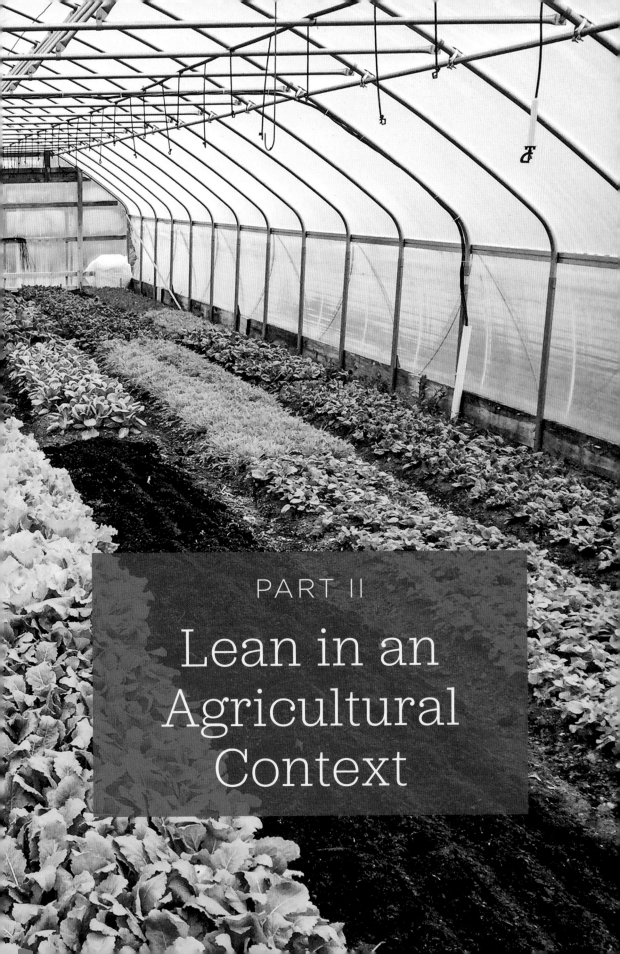

PART II

Lean in an
Agricultural
Context

CHAPTER 11

The Lean Farm Start-Up

*There is no shortcut to achievement. Life requires
thorough preparation—veneer isn't worth anything.*

—George Washington Carver

On older and more established farms, once wasteful habits have created inroads and taken hold they require a lot of work to root out. A new farmer has a special opportunity to establish lean habits from the beginning and avoid years of lost effort.

Learning by doing is a lean principle.

Our own start-up was not perfect. We made our share of mistakes. I could fill pages describing crop failures and construction fiascos, overambitious schemes and missed opportunities. Yet we did stick to a set of start-up rules, which I describe below, that helped us avert disaster and turn profits early. The rules are common in lean enterprises, though we were unfamiliar with lean ideas at the time. If we had had a list like this one, we might have adhered to the rules even more closely and avoided a few more hiccups. Other thriving small farms I have visited followed these rules, too. Those aspiring to farm on a larger scale with more capital investment will also find useful information below, though the ideas are aimed at first-time farmers starting out with little cash and hoping to make a living on just a few acres.

Principle 1: Put in your 10,000 hours (develop personal capacity first).
Principle 2: Test in small batches.
Principle 3: Add infrastructure capacity in small increments.
Principle 4: Avoid bad debt.

Principle 1: Put in Your 10,000 Hours (Develop Personal Capacity First)

Many farming entrepreneurs assume that farms are like stock market accounts: invest a truckload of capital in land, buildings, and equipment, and money will magically start flowing. But farming is a highly skilled job, not a financial game.

In his book *Outliers: The Story of Success*, Malcolm Gladwell makes the case that to become masterful at any task requires 10,000 hours of practice, or about ten years of dedication to your field. Elite chess players compete for ten years before they win breakthrough tournaments (with the exception of Bobby Fischer, who took nine years). Mozart famously started writing music at age five, but he didn't produce masterpieces for another twenty years. Between 1961 and 1962, as a new band, the Beatles performed for 270 nights—eight hours per night—in less than eighteen months. They put in their hours! Bill Gates started programming as an eighth grader in 1968. He had practiced well more than 10,000 hours before he dropped out of Harvard to start his own software company. Time and again, "10,000 hours is the magic number of greatness,"[1] says Gladwell.

Mastery in farming is no different. Only after ten years of farming for a living do I feel like I have a handle on the trade—and I grew up on a farm. Don't be discouraged. You can start your farming enterprise before you've clocked 10,000 hours of practice. But be realistic. Mastery will follow years of dedication.

The Value of Manual Labor

Much of your 10,000 hours should be manual work. There is nothing wrong with a university education in agriculture; it can give you a good foundation. But learning with your hands is a must in order to farm well. Farming is a tactile profession as well as a mental challenge. Skilled farmers who rely on their farms for income will give you perspective and knowledge you won't find anywhere else. At some point in your education, get on a farm! Be choosy. Not all farms are alike, and not all farmers share the same set of capacities. Get a sense for the type of farm you'd like, then approach a farmer whose real farm matches the one you dream of. The best university programs in farming will incorporate manual labor and on-farm internships with sit-down teaching.

Our society by and large does not value manual labor. We have this "idea that work is beneath human dignity, particularly any form of manual work," as Wendell Berry writes. As such,

We have made it our overriding ambition to escape work, and as a consequence have debased work until it is only fit to escape from. . . . But is work something that we have a right to escape? And can we escape it with impunity? We are probably the first entire people ever to think so. All the ancient wisdom that has come down

Transplanting tomatoes.

to us counsels otherwise. It tells us that work is necessary to us, as much a part of our condition as mortality.[2]

We'd rather think than apply our thoughts. Many aspiring farmers did not grow up in communities where manual labor was valued, and so they often lack basic skills—and necessary attitudes about work—to farm well.

Manual work is not all drudgery. It connects us with the physical world; it grounds us. Its returns are physical health and a right sense of place. Manual labor can't be replicated by reading a book. Its joys can't be experienced behind a screen. My father used to pound his fists together in the middle of his farmwork, furrow his eyebrows, and yell with a grimace (and half a smile), "Hard, physical labor!" It wasn't a complaint; it was satisfaction.

Farming is a *manually* applied science. I love learning about plant varieties, their habits and origins. I am interested in a plant's natural (preferred) environment and what temperature a particular seed needs in order to germinate. Farming is taking this knowledge and applying it with your hands to produce food. To grow a tomato, I combine what I know about botany and greenhouses and heating systems and soils to create an ideal environment for a lifeless seed to turn into a plump, red fruit. The ideas start in my head, but I nurse the tomato along with my hands.

It takes doctors at least eight years of study and practice to become licensed. Then they continue to learn on the job. The same amount of work is required of master plumbers, electricians, and carpenters. Likewise, to gain mastery, farmers must amass years' worth of knowledge about botany, biology, soils, plant disease, produce harvesting, post-harvest techniques, and produce storage, not to mention sales, marketing, business management, staff management, and accounting. All this in addition to "trade" skills like tractor operation and maintenance, small-engine repair, engineering, construction, basic plumbing, electricity, and perhaps heating and ventilation as well.

Principle 2: Test in Small Batches

To find new markets test, test, test. And keep your tests small. Farming is risky enough as it is; don't add more risk by overspeculating.

If you want to raise bees for wholesale distribution, produce just enough honey to let your potential customers get a sense of your product and how they might use it to make money in their businesses. Then you can ask them to project the volume they might order and the precise type of products and services they want, that is, how they want it packaged and when they want it delivered—and only then can you begin forecasting the right volume to start producing.

With this approach your customer is your partner. Together you develop your product. Explain your goals and tell them that with their help you can grow to meet their demand. It's okay for your supply to be somewhat erratic until you figure out precisely what customers value and how best to scale up and smooth out production. If you take careful notes that first season, you should by the end of the year have a powerful blueprint for very lean production in the second year. This is how new farm ventures can take off very quickly.

Software and Internet start-ups commonly use small-batch testing, coupled with a heavy dose of customer feedback. Instead of investing thousands of dollars in a finished product, these high-tech entrepreneurs simply get their bare-bones products into the marketplace and then fine-tune them based on feedback from real customers rather than mere speculation. Hence the ubiquity of feedback surveys on just about every website you visit. Surveys are the mechanism by which you, the customer, guide the development of a product. Eric Ries discusses this approach in *The Lean Startup*.

The cost savings is tremendous when you let the customer shape your product. You remove risk and give yourself time to precisely define value. Otherwise, if you are guessing at what might sell, you will likely lose years if not decades going down rabbit holes.

Principle 3: Add Infrastructure Capacity in Small Increments

Too many new farmers start out at a sprint, building infrastructure capacity at an alarming rate. They buy land, tractors, and the latest and greatest tools. They put up buildings and greenhouses. They add lanes and assemble a fleet of top-of-the-line delivery vehicles. Their infrastructure capacity is out of balance with their personal capacity.

In contrast, the lean approach is to add capacity in small doses, as you need it, as you're capable of using it well. In the very beginning, how much infrastructure capacity do you need? Lean says very little. Start out with only what you need to produce your sampling. When you have a solid plan to produce and sell more, then scale up. It sounds simple, but many lack the discipline to grow at the right pace.

In other words, don't jump to expensive equipment solutions before working out process kinks first. Start with a small greenhouse, master its use, and then build another. Get by with pushcarts and walk-behind tillers and master your growing before you buy tractors and wagons. Deliver your food in your car or truck, establish solid accounts, and then buy that delivery vehicle. With our mushroom experimenting, we could spend thousands of dollars and build the perfect grow house, complete with state-of-the-art climate-control systems. Perhaps someday we'll get there. But until then our goal is to get a couple of mushroom varieties to the marketplace with minimal expense. For now, this means growing in a small area in an unused milk house, using heaters from our stash of stored prototyping supplies, and staying within a budget of less than $1,000.

An oyster mushroom trial.

Principle 4: Avoid Bad Debt

To put process solutions before equipment solutions does not mean you should never invest in your farm or take out loans. You need a certain amount of investment—infrastructure, land, and equipment—to get off the ground and to grow. We were helped by bank and family loans to help us purchase our property, build greenhouses as we needed them, and buy our small tractor once we had outgrown our tiller. And we worked off the farm for several years to subsidize our start.

However, lean farming asks that you distinguish good debt from bad debt. The debt we took on was good debt because it was applied to an investment that we were confident would pay itself off in a reasonable amount of time and yield returns for many years afterward. Bad debt is speculative investment. Taking on bad debt means spending money you don't have to finance a scheme you haven't market-tested and that could easily flop. You might need financing to get going and occasionally to scale up a project, but don't get carried away. Debt that the farm can't pay off will kick you off the farm. You'll be forced to work for someone else so you can service your loan.

Farming costs money, and there is no way to avoid risk. Land prices are steep in many areas, and you need buildings and equipment to produce at a volume that can sustain even a small enterprise. These lean principles will help you keep your debt load sane and remove a lot of risk from the bumpy first few years.

CHAPTER 12

The Limits of Lean in Agriculture

The farmer lives and works in the meeting place of nature and the human economy. Farmers either fit their farming to their farms, conform to the laws of nature, and keep natural powers and services intact—or they do not.

—Wendell Berry

Can a system that works well for building cars really apply to farming? Are there not fundamental differences between factories and farms that should keep us from learning from one another? What about sustainability? Can we be too ruthless in our efforts to cut costs?

From our fields we see contradictions every day. The Amish farmers on either side of us might be plowing with horses or baling hay with their families, using old but well-tended equipment, while across the road a large-scale "English" farmer plants hundreds of acres of corn in just a few hours from the comfortable seat of a GPS-controlled tractor. A few miles down the road, dozens of factories at an industrial park mold steel, aluminum, and copper into manufactured goods.

Our farm is located in the center of Elkhart County, Indiana, home to hundreds of small and large farms that produce milk, eggs, cheese, fruits and vegetables, grains, and more for corporate buyers like Whole Foods and Cargill, as well as for neighborhood markets, roadside stands, and

Farms are unique production environments because living nature is the core ingredient of farming. Photo by Emma Gerigscott/ Clay Bottom Farm.

food co-ops. The county is also a manufacturing hub. Every year tens of thousands of RVs are built within the 30-mile radius encircling our farm.

In the Introduction, I discussed how farms are special places because their raw materials are living, not inert. In this chapter I discuss three models of production—models that apply to both farms and factories—as a way to further explore this difference. Then I'll give examples of agribusiness interests applying lean with disregard for this distinction.

Three Models of Production: Mass, Craft, and Lean

In *The Machine That Changed the World*, James Womack, Daniel Jones, and Daniel Roos identify two basic models that humans have devised to make things: mass production and craft production.[1] Lean is a third option.

MASS PRODUCTION

In mass production the goal is to create standardized goods in high volume. Production relies on expensive and single-purpose machines. While business owners might have sophisticated knowledge, workers are often narrowly skilled or unskilled, completing tasks that are simple and repetitive, as in the case of workers on many assembly lines. The model is resource-intense, gobbling raw materials rather than rationing them. According to Womack, Jones, and Roos, "Because the machinery costs so much and is so intolerant of disruption, the mass producer adds many buffers—extra supplies, extra workers, and extra space—to assure smooth production."[2] Mass producers earn their profits through massive scale rather than through flexibility or smooth process. By necessity, the variety they offer is low.

Mass production in agriculture takes the shape of large-scale farms specializing in just a few crops or animals. Work is often monotonous, as when pickers on sprawling fruit and vegetable farms spend weeks on end performing simple harvesting tasks. These operations are called industrial farms because they more closely resemble mass-production industry than they do traditional farming. On the surface, it would appear that this type of agriculture benefits from an economy of scale. For example, fields in northern Indiana, where we live, are usually filled with a sea of either corn or soybeans. On our own farm, we spend several hours per week year-round harvesting more than forty different types of crops from our plots. Field-crop farmers can accomplish their single-crop harvests in a

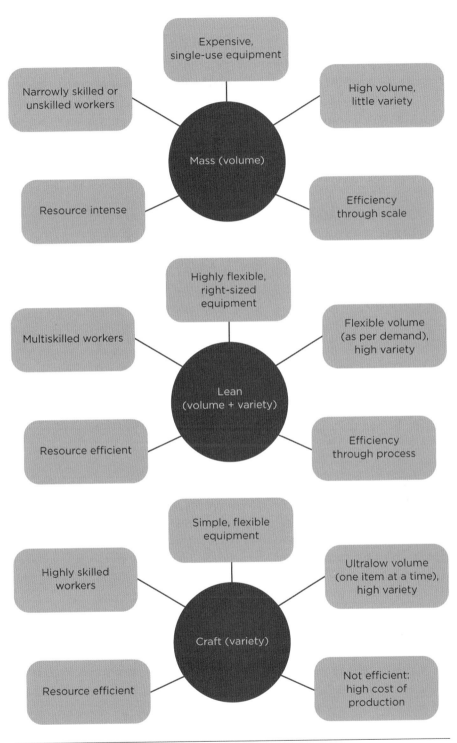

3 Types of Production. Mass production specializes in volume; craft production specializes in variety. Lean combines the best of both. Based in part on concepts from *The Machine That Changed the World* by James P. Womack, Daniel T. Jones, and Daniel Roos.

matter of a few days, if not hours. It would seem that they have a corner on efficiency and value.

However, as I explained in Chapter 10, it is a myth that large-volume work is always more efficient than smaller-volume work. Mass production contains costs that are not so easy to see. For example, commodity-crop growers cannot quickly alter what they are producing to respond to changing market conditions. Their equipment is dedicated to a single purpose, too specialized and expensive to allow them to produce crops other than the one or two their equipment is designed to handle. They lose out when markets shift out of their favor.

Also, their fields sit empty much of the year because their scale is too large for them to maximize the investments they have made in their land. Aside from a few weeks in the spring and fall, their costly equipment likewise sits idle most of the time, depreciating in value.

Henry Ford's Model T was one of the first mass-produced consumer goods. The Model T was not a custom-built automobile. Ford famously said that his customers could have any color they wanted, "so long as it is black." When a product is in high demand for a long time, such as the Model T was in the 1930s, there is no reason to customize. Ford wasn't set up for change because his assembly line relied on specialized equipment designed to crank out standardized parts at high volume. As the market changed, the Model T could not change with it and was finally phased out.

CRAFT PRODUCTION

Craft production is just the opposite of mass production. The craft maker is interested in producing the exact good that a customer wants. Often there is a close relationship between maker and a buyer, such that the buyer can customize orders and the artisan can produce precisely what the buyer demands. Variety is paramount. Workers are highly skilled and use simple but flexible equipment. Volume is very low—one item at a time—and because production is labor-intense, the cost of production is high. Since work is calibrated and fine-tuned, resources are used efficiently. Historically, all production prior to the early 20th century, that of shoemakers and blacksmiths, for instance, was small-volume craft production. Today a custom cabinetmaker engages in craft production.

Before grains were commoditized through the Chicago Board of Trade's grading system in the 1850s, American farms fit a craft production model, with farmers raising a wide variety of crops. Farm goods were attached to real farmers who sold them directly, without the go-between of the commodity market. Farmers took pride in quality, and one farm often offered an enormous selection of goods, from pork, chicken, beef, and milk to water-melons, green beans, tomatoes, and wheat. When grain crops became commodities, they could be traded as generic items— number 2 corn—rather than as products linked to specific farms. Farmers were freed to pursue the mass production of one or two items.[3]

An obvious drawback for craft producers is that specialization often increases costs, which increases prices, and higher prices mean fewer customers. Also, because one-of-a-kind production requires the attention of a single artisan or a small group of highly trained artisans, production volume is often too low for such businesses to remain viable in today's economic environment, although new Internet markets like Etsy and eBay do make it more possible for artisan producers to reach a larger customer base. In farming, food hubs and farmers' markets offer similar benefits to craft farmers.

LEAN PRODUCTION

Lean seeks to combine the best of both mass production and craft produc-tion. A lean enterprise seeks to cut costs and produce a measurable volume while staying flexible enough to respond quickly to the marketplace and offer customers more choice. Womack, Jones, and Roos explain it this way: "Lean producers employ teams of multiskilled workers at all levels of the organization and use highly flexible . . . machines to produce volumes of

Demand changes constantly, so lean farming relies on flexible production and multi-purpose tools.

products in enormous variety."[4] Resource use is wise—just enough—not careless. Lean producers use smart process rather than enormous scale to achieve efficiency and generate profits. Workers can perform many complex tasks and switch tasks quickly to meet changing demands.

As this book shows, we have tried to adopt this model on our farm. Through the selective use of machines (right-sized technology), we aim to grow and sell in a volume that can support us and at the same time remain highly flexible in what we grow in order to respond to changing demands for food in our community. For example, we use a small but efficient seeder that sows a range of more than thirty different crops that we produce. We use the same tractor and bed shaper setup seen on commercial farms in California, though our equipment is much smaller so we can squeeze them into tight corners and maximize our land use. Similarly, our custom-built root-digging implement can be found in larger form on commercial farms. We grow efficiently, yet we can swap out equipment and change what we grow as well.

US Farming: A Rut of Mass Production

Most large-scale US farming is classic mass production. Corn and soybeans, which rely on expensive, single-use machines, are chronically overproduced, resulting in wild price fluctuations, even with a massive subsidy system (a $5 billion farm bill) designed to counteract the problem. Even in years when production more closely matches demand, the phenomenon is short-lived, as producers in Russia, Eastern Europe, South America, and other parts of the world quickly flood the market. The system is resource-intense: US agriculture is the single largest consumer of fossil-fuel energy compared to any other industry.

In response, rather than change this system, the modus operandi of US agribusiness and their Wall Street and government backers is to do more of the same. Instead of motivating farmers to diversify their products and seek out new markets, our system rewards their overproduction, perpetually masking the problem by pushing the excess onto the public in the guise of ethanol, corn-derived plastics, processed foods, corn syrup, food preservatives, food coloring, emulsifiers in beauty products and antifreeze, and even vitamin supplements (most vitamin C used to fortify foods is synthesized from corn, not naturally occurring). Simply put, we are drowning in corn and soybeans.

Since competitive marketplaces don't tolerate stagnation, most other industries have by now had to adapt and align their products more closely with what people actually want. But because of rooted interests such as the packaged food and ethanol industries, farming has not changed.

Of course, food waste and obesity are two results—our food system is set up so that overproduction is part of the plan. Our diets are filled—indeed, stuffed—with corn oil, high-fructose corn syrup, and countless other outputs of the corn and soybean industry. Efforts to address food waste must contend with these production realities and not just focus on consumer habits, although getting consumers to diversify their palates is part of the solution.

You might think that the excesses of mass-production farming are so obvious that the industry would be scrambling for ways to lean up, that is, to encourage diverse production and more community-driven farming. Likewise, you would think the next wave in agricultural engineering would be inexpensive multifunction equipment that can quickly change from producing one type of crop to another. Instead, at the 2014 farm show in Fort Wayne, Indiana, all eyes were on the driverless tractor, an "unmanned ground vehicle" costing several hundred thousand dollars with one purpose in mind: to haul even more grain wagons full of corn and soybeans out of the breadbasket of the United States.

A final note about mass-production agriculture: I say US *farming*, not the US *farmer*, is stuck in a rut: farmers—even mass-production farmers—are smart people who by and large are doing what they can to save their farms. Farmers who produce just one or two crops each year do so because the country's system of agriculture rewards them for doing so and, through taxpayer subsidies, protects them to some degree from overproduction.

The entire enterprise of agriculture needs to adopt a waste-free approach that combines the volume of mass production with the customer focus and flexibility of craft production. This approach would stop pushing excess supply and instead put people at the center of our food system, so that what local communities actually want to eat would be the determining factor pulling crops from surrounding farms.

Some argue that in fact our communities *want* the industrial food products—the Twinkies and Pop-Tarts—that line grocery store shelves. Don't be fooled. The agribusinesses that benefit from our gluttony spend billions of dollars every year convincing people to make bad food choices. And their government backers chip in billions more by subsidizing these foods so that good food choices—like farm-direct eggs, milk, meat, and produce—are put out of financial reach for a giant swath of eaters. The industrial agricultural

Wendell Berry proposes that farmers work at the intersection of nature and the human economy. Photo courtesy of David Johnson Photography.

machine has mastered the art of salesmanship, astutely defined by Wendell Berry as "the craft of persuading people to buy what they do not need, and do not want, for more than it is worth."[5]

Why Farms Are Special Places

The primary way mass-production agriculture fails us is not that it produces too much of the same thing, destructive as that is. Rather, it is that it treats farming like any other industrial business. It ignores nature, which is the unique environment of the farm.

The fundamental difference between factories and farms, between manufacturing and agriculture, is that farm products are or once were *alive*, meaning sentient, mysterious, and worthy of our ethical consideration. None of this is news to traditional farmers who farm without chemicals and with high regard for nature. They know through experience that plants, animals, and even the soil are deeply mysterious and full of life and that respect for this fact precedes true productivity.

So there is a moral element to farming. As I've mentioned, farmers have a unique relationship with the living parts of their enterprises—with nature—that does not exist on the assembly line. We are as chess players constantly adapting to a changing board. Farmers depend on the well-being of nature, and the well-being of nature depends on farmers practicing their trade with care.

This is not to say that manufacturing is or should be disconnected from nature. The raw materials we use to build cars and airplanes and cell phones don't fall out of the sky; they are plucked from the earth, and not without consequences to people and places. There are moral elements there, too. But farmers have a more obvious and direct connection. Their day-to-day activities are part of a

Nature—the raw material of farming—is alive, mysterious, and constantly changing. Photo courtesy of David Johnson Photography.

living cycle that ultimately envelops human beings, through the intimate medium of food, into the cycle of life.

Lean Overreached

On farms, waste elimination is not always positive. The tendency to overreach for profit is especially pronounced when lean is applied recklessly to mass-production farming. The people who bankroll and often own gigantic farms are frequently not farmers but investors who live far away from the ecosystem of the actual farm. There is a wide disconnect. Some specific cases illustrate where lean tools don't always fit so neatly in the living system of agriculture.

TOXIC WATER: STANDARDIZATION TAKEN TOO FAR

In an effort to create smooth flow on the production line and ensure that untrained workers can get on board quickly, lean managers employ standardization. As I showed earlier, we have standardized the totes we use, our harvesting practices, our greenhouse ventilation, and more. Standardization has saved us countless hours and thousands of dollars of lost time. But when standardization is taken to mean you should grow only the one or two crops that yield the highest return, then the environment and, eventually, humans suffer.

Consider the water crisis in Toledo, Ohio, in 2014, in which half a million people were left without clean drinking water for weeks because of toxic contamination. The root cause of the ongoing crisis was overuse and underregulation of phosphorus in the form of chemical fertilizer and manure applied to surrounding farmland. While there are many sources of phosphorus flowing into Lake Erie and Ohio's mainland lakes, the dominant stream is agricultural runoff. From a lean standpoint, then, Toledo's water crisis can be explained as a case of overstandardization.

Even though the application of chemical phosphorus may be a great way to boost plant growth, too much of the same thing is wreaking havoc. Industrial farmers refuse to destandardize fertility practices and try a more diverse approach to land management by, say, using crop rotation, composting, and cover crops to build healthy soils.

Water pollution is not limited to Toledo, of course. It is a national problem. According to a report from the Environment America Research and Policy Center, 206 million pounds of toxic chemicals like phosphorus were dumped into US waterways in 2012. The worst polluting state was my own, Indiana, followed by Texas, Louisiana, Alabama, Virginia, Nebraska, Pennsylvania, Georgia, North Carolina, and Ohio.[6] Change does not appear on the horizon. The 2014 US spending bill, drafted just months after the Toledo crisis, requires the Environmental Protection Agency (EPA) "to withdraw a new rule defining how the Clean Water Act applies to certain agricultural conservation practices." Further, "It prevents the Army Corps of Engineers from regulating farm ponds and irrigation ditches under the Clean Water Act."[7]

ANIMAL CONFINEMENT: MAXIMIZING FIXED COSTS TAKEN TOO FAR

Fixed costs, the investments you've made in your enterprise that you expect to last beyond several production cycles, can be *over*maximized.

On many farms maximizing fixed costs takes the form of efficient use of space. But it can turn into abuse of animals when farmers confine ever more animals into tighter and tighter spaces. In industrial hog production, for example, gestational crates, or sow stalls, house sows during pregnancy, which amounts to most of the pig's life. The crates allow for a large number of animals to fit into very tight quarters, and on some farms the cement-floored crates are so small that the sows don't even have room to turn around. Six major companies, including McDonald's, General Mills, and Hillshire Farms, have asked their producers to phase out the use of such confinement practices.

RBST AND METHANE POLLUTION: CYCLE-TIME REDUCTION TAKEN TOO FAR

In farming, cycle time, the time to complete one cycle of production from start to finish—depends a lot on natural forces. In fact, it's largely out of a farmer's hands: there is little a farmer can or should do to speed the natural growth cycle of plants and animals.

But oh, how we try! Animal farmers in the United States are injecting Monsanto's bovine growth hormone, or rBST, into their animals at alarming rates to make them gain weight faster, thus shortening the cycle time of meat production. Hormones also boost milk production. But in humans they are associated with increased rates of cancer, especially breast and prostate cancer.[8] Studies also suggest that the drug could lead to allergies.[9] And there is debate about the role of rBST in the early onset of puberty. Likewise, plant breeders are increasingly resorting to genetic modification to develop plants that are bigger, stronger, and faster.

Gas pollution from cattle is a by-product of cycle-time reduction. According to a United Nations report, cattle rearing generates more greenhouse gases than does transportation, accounting for 18 percent of greenhouse gas emissions measured in CO_2 equivalent.[10] Primarily, the report says, 37 percent of all human-induced methane (twenty-three times as warming as CO_2) is "largely produced by the digestive system of ruminants."[11] As such, livestock represent one of the "top two or three contributors to the most serious environmental problems, at every scale from local to global." The report warns that the "impact is so significant that it needs to be addressed with urgency."[12] The EPA confirms that global agriculture is the primary source of methane emissions.[13]

The root of the problem is diet. In an effort to speed up growth, farmers feed grains to animals whose digestive systems were never meant to handle such food. What's needed, says the UN, is "improved animal diets to reduce enteric fermentation and consequent methane emissions."[14] In other words, livestock, like many humans, are consuming too much corn.

Again, industrial agriculture shows no sign of change, and the government seems happy to collude. According to the *New York Times*, the 2014 spending bill dictates that the government cannot require farmers to report greenhouse gas emissions from manure management systems. And it stipulates that ranchers are exempt from purchasing greenhouse gas permits for methane emissions produced by "bovine flatulence or belching."[15]

As these overreaches show, there should be limits as to how the lean system is applied in farming. Without limits, cutting costs can go too far, resulting in farming that does not honor nature. When farmers respect limits, however, their farming can have incredibly positive impacts, as the next chapter will show.

CHAPTER 13

Lean for More Than Profit

Cultivators of the earth are the most valuable citizens.
They are the most vigorous, the most independent, the most
virtuous, and they are tied to their country and wedded
to its liberty and interests by the most lasting bonds.

—Thomas Jefferson

The most meaningful walk I've taken around our farm was a few months before this writing, on the day our son, Arlo, was born. We were lucky to have a home birth, and we wanted to show him the world he was born into. So after his first long nap, and after the midwife and family all left, we swaddled him up and started a slow walk.

We started by showing him our beds of red and green and yellow head lettuce. Then we passed the stalks of ruffled kale, woody pepper plants, and fig trees dotted with pear-shaped fruits. We had to wonder what it must be like to discover illuminated nature for the first time, after hearing only muffled noises for so long. Every day since he was born, Arlo has touched new textures—the waxy coating of an aloe vera plant, the roughness of a maple branch. He smells new scents, like chopped garlic or fresh-cut parsley. He sees new shapes: jagged stones, slender blades of grass, and arching trees. The farm and the world are his to discover.

Last year I helped out at a food pantry farm. I had loaned the group a few of our tools and showed them how to plant beans, peppers, and tomatoes. I worked with one young girl, probably eleven or twelve years old, who was from an urban neighborhood and had never done any sort of gardening. As we set a pepper transplant into the ground, she told me, "Ben, this is *so* cool! I had no idea where peppers came from. They just always showed up at the store."

Farms are special places that connect us to what we eat and to lively and mysterious nature. We all should cherish farms.

From the viewpoint of a farmer and now a father, my main concern with farming that exploits nature is that such farming leaves farms and rural places uninteresting. What child would choose to explore a 10,000-acre monocrop field, or a cement-walled confined animal operation, over a 1-acre patch containing ponds, trees, animals on open pastures, and all manner of fruit and vegetable plants?

Overreaches like those I discussed in the previous chapter pose a challenge to those who use lean: ponder deeply your primary motivation. If your drive is profit to the exclusion of quality of life, then lean can deliver to you, too. But sooner or later you will find yourself exploiting land and animals to take care of your wallet and outside investors. There is another option. You can use lean as a tool to further a more holistic cause, to restore the earth and connect your farm—your slice of nature—with children and other eaters in your community. In that case, your primary concern, as Wendell Berry says, is to nurture. As Berry writes,

> *The exploiter's goal is money, profit; the nurturer's goal is health—his land's health, his own, his family's, his community's, his country's. Whereas the exploiter asks of a piece of land only how much and how quickly it can be made to produce, the nurturer asks a question that is much more complex and difficult: What is its carrying capacity? (That is: How much can be taken from it without diminishing it? What can it produce dependably for an infinite amount of time?). . . . The exploiter typically serves an institution or organization; the nurturer serves land, household, community, place.*[1]

Nurturing agriculture is an older tendency than exploitative agriculture. It is "the tendency to stay put, to say, 'No further. This is the place,'" as Berry says. This was the tendency of indigenous peoples and of the noble small farmers—the "cultivators of the earth"—that Thomas Jefferson lauded as our country's most valuable citizens. Profit is not eliminated as a concern for nurture farmers. "The nurturer expects to have a decent living from his work," says Berry, but his primary motivation is "to work as *well* as possible."[2]

To accomplish this twofold feat—to work well *and* earn a decent living—is an undeniable challenge in our current economic environment, where corporate interests have put farming out of reach for many young people. It requires a farmer to bypass the industrialized system of growing and selling food. Farmers must be excellent at their craft but also independent and shrewd—both good farmers and good businesspeople. This challenge compelled us to take up our study of lean. After we saw what this system from Japan could do for our small farm, I felt compelled to write

Lean on farms has many uses beyond profit.

this book. I hope to teach how to farm shrewdly and with as much business wit as your corporate competition so you can sell your food independently of their scheme. With the right motivation, leaning up can afford a place where you are able to work *well*, contributing to the health of your household and to the ecosystem of the farm while remaining viable.

To work well also means solving problems beyond the farm. Lean tools can help. In the previous chapter I examined cases of overreach, where farms are exploited blindly for profit alone. In this chapter I suggest specific ways that lean might be of use in the other direction, to further causes even beyond profit—specifically, to bolster a farmer's efforts to help solve five particular problems in the food system today.

1. Lean to Eliminate Food Waste

According to a US Department of Agriculture study released in February 2014, "31 percent—or 133 billion pounds—of the 430 billion pounds of the available food supply at the retail and consumer levels in 2010 went uneaten." The retail value of this food was a staggering $161.6 billion, and the estimated calories were 141 trillion, or 1,249 calories per person per day.[3]

There is food waste on the farm, too. A Natural Resources Defense Council (NRDC) paper on food loss estimated that 20 percent of fresh fruits and vegetables never leaves the field.[4] These excesses, dubbed "walk-by losses," can occur because of weather or insect damage, market volatility, and overproduction. The NRDC estimates total food loss—including losses from storage, transportation, and retail and consumer loss—at 40 percent. The consequences are many: "Getting food to our tables eats up 10 percent of the total U.S. energy budget, uses 50 percent of U.S. land, and swallows 80 percent of freshwater consumed in the United States. . . . Moreover, almost all of that uneaten food ends up rotting in landfills where organic matter accounts for 16 percent of U.S. methane emissions."[5]

Lean tools can help farmers do their part to reduce food waste. A better focus on customer value will ensure that farmers are growing what customers will buy, not just what farmers want to grow. Better forecasting and production control will help farmers plant the right amount and not too much. Smoother flow in the field and in post-harvest handling can minimize defect and loss.

The Arnolds at Pleasant Valley Farm in upstate New York are proud of the work they've done to cut food waste. Paul Arnold smiled when he told me, "We don't even produce enough waste to feed our two hogs." One option to help lower waste on your farm is to track food waste as a metric: add up the weight or the value of composted food every week and find ways to make the number go down. This is an example of a metric that should excite farmers and workers alike and benefit your farm's bottom line as well as the environment. Some food losses, however, are inevitable. Weather, pests, and disease will always find ways to foil our crops or cause losses on animal farms. It's impossible to expect perfect crops out of every harvest. But leaning up can keep those losses small.

2. Lean to Strengthen Local Foodsheds

A foodshed is the geographic region that produces the food for a particular population. More communities are realizing that many farms in their regions are not producing food for local people. These farms might be *in* the foodshed, but they are not *of* it. Their food is shipped away, to be packaged into processed foods or sold as ethanol. This leaves communities vulnerable. If the global food system, which makes such farming possible, is disrupted, where will food come from? Any number of factors—from weather disruptions to stock market collapse to a spike in fuel costs—could quickly and drastically alter the industrial food chain. So foodshed builders aim to persuade producers to keep food local and to connect buyers with farms that take up the challenge.

Farmers can do their part. Lean doesn't necessarily mean going to all lengths—and traveling all distances—to find the highest-paying markets. Lean tools for growth, like putting the customer first and cutting costs, can help a farm succeed in local markets as well.

If profit were our only motive, we might drive our food to higher-paying but more distant markets. However, we want to be part of the local food movement. We value our small, local farmers' market as a community gathering place and a centerpiece of our local foodshed. We want to make sure our food is available there. Every so often, a grateful customer will approach us and say,

Farmers can use lean to strengthen local foodsheds. Photo courtesy of David Johnson Photography.

City Dwellers and the Foodshed

Just as farmers can take steps that strengthen local foodsheds, so those who live in cities have a role to play as well.

Urban places might feel disconnected from rural places, but cities and farms are tied inextricably together. If you are a city dweller, every time you eat—unless your meal was grown in your backyard—you connect yourself to a rural place. The stores that sell food are as much a part of farming as the land on which food grows. Food buyers drive the decisions farmers make about what to plant and what practices to put in place. The water farmers use to irrigate is the same water that will end up, after it is filtered through fields and carried through streams, in the well from which cities drink, as the people of Toledo know well. The air farmers purify or pollute is the same air city people will breathe. The land farmers use to grow food is essential to farmers economically—they need good land to make a living—but essential to all of us nutritionally: we all need good, productive land to survive.

Urban eaters must become essential partners in restoring intention to farming. To eat local food is to make a special and conscious choice. You are joining forces with local farmers to take back control over a food system that corporate interests have co-opted. In the process you restore a piece of earth as you and your farmer undo the degradation on a slice of land. You increase your food security because you help to build an alternative and independent food system whose survival doesn't depend on the interests of a wealthy few who might live thousands of miles away. Rather, you will have food, and plenty of it, for as long as the ties between farmers and eaters remain strong.

To support a local foodshed is really that simple: eat local food. Don't be fooled—millions of dollars are spent every year trying to confuse the issue of what it means to support farms. The US farm bill, now a nearly $5 billion package that passes Congress every five years, is an example of legislation that contains a slew of measures to appease corporate interests in the name of "supporting America's farms." According to the *Economist*, 75 percent of subsidies in the farm bill go to the wealthiest 10 percent of farms, not small farmers selling at your local farmers' market but millionaire-owned corporate enterprises. These aren't small subsidies. The mean subsidy received by the top 1 percent of policyholders was $227,000. The mean subsidy for the bottom 80 percent was $5,000.[6] There is a difference between supporting farms that are good for communities and subsidizing industrial intrusion into our rural places.

"Thanks for growing our food. Please don't quit." Comments like these keep us motivated to sell locally. To strengthen foodsheds, it is important to keep locally owned food institutions thriving. When people ask me what they can do to support local farmers beyond buying their food, I tell them, "Volunteer at your farmers' market or serve on the board of your food co-op. Keep those institutions strong; your farmers need them!" For their part, farmers can keep those institutions strong by selling there.

As of this writing, 90 percent of the food we grow on our farm ends up on plates within 10 miles of the farm. We use lean to keep our farm both profitable and local.

3. Lean to Provide Local Food for Those Who Can't Afford It

Another reason we sell through our farmers' market is that we are able to partner with other farmers and government programs to get our food into the hands of those who normally couldn't buy it. Every year, the state of Indiana issues Women, Infants, and Children (WIC) vouchers to low-income mothers for use only at farmers' markets. The program ensures participants receive fresh, healthy food and supports local agriculture at the same time. In addition, some churches in our area provide prepaid farmers' market baskets for community members to use. The community members fill their baskets at any vendor's booth. The farmers' market itself operates a Share the Bounty program in which churches, organization, and individuals donate to a fund that is used to provide farmers' market vouchers to families and individuals. The fund also enables the market to match up to $10 worth of electronic benefit transfer (EBT) withdrawals of US government-issued SNAP (food stamp) money. It's much more efficient for us to work with established programs like these rather than to tackle problems like hunger on our own. Farmers who care about these issues can use a metric that says a certain percentage of their food should go to low-income community members, then use a lean approach to make sure their efforts are well spent.

There is no rule that says only for-profit farms benefit from a lean approach. Community and food bank gardens can expand their efforts with lean tools as well. Even service organizations, like food kitchens and food pantries, can apply lean lessons to maximize their work. For example, food pantries in our area recently coordinated with one another to establish Seed to Feed gardens: a coordinator forecasts the amount of produce needed—and when it is most needed—and growers and volunteers use the information to plan out gardens that supply produce to a centralized facility. The produce is then distributed to pantries across the county according to their needs. The coordination is an example of production control; it cuts enormous waste and maximizes everyone's efforts.

4. Lean to Reduce Farm Trash

The detritus of farms—containers, plastic films, used tires—fill up landfills at an alarming rate. It shouldn't take so much waste to grow our food.

Nature provides a good example of a lean system. Last spring, in the rafters of a barn loft that we have cleaned out to host farm-to-table meals, a pair of swallows built a nest. When it came time to host a meal, we didn't have the heart to remove the nest but just cleaned up the droppings and made sure tables were not directly under the nest. The swallows flew in and out while diners ate, becoming a topic of dinner conversation. We watched them build with creative bits of leftovers—a string of plastic from landscaping fabric, bits of straw, twigs, and blades of grass. In the wild, everything is repurposed. Bird nests are not excessive, and they do not generate waste. They are the right size to serve their purpose. Human houses, on the other hand, can be grotesquely large, and many construction sites generate thousands of cubic yards of waste.

A cleaned-up farm is inviting for visitors.

Farmers would do well to take note of excesses. Lean strategies, like keeping on hand just enough tools and supplies, can help reduce trash. When you set specific targets, you waste less. If you must use landfill-destined supplies, consider ways they might be repurposed on your farm or somewhere else first. This is one way to lean up for the sake of the environment, not just for profit.

5. Lean to Create Inviting Farms

Children in urban communities need farms to roam around on, but there aren't enough of them. Lean can clean up a property and make it more welcoming. On our farm, messes still crop up, but visitors can usually walk through rows of tomatoes and peppers or inspect our microgreens greenhouse without tripping over a hose or getting bound up in debris. Lean tools like 5S sorting and short, high-frequency cleaning keep things tidy. For children, our farm is more pleasant as well.

A nephew and niece playing in a greenhouse.

One day we were watching a friend's son, Adam, who was eight years old. For several minutes, Adam, working on his knees, helped me harvest beets. I pulled them and cut the tops, and he put them in a crate. Then a tiny frog hopped out from behind a beet green. Suddenly, Adam couldn't care less about beets. He got up and followed the frog around, studying its habits, completely captivated by every move. If you wonder what it means to create an inviting farm, I suggest a simple rule: observe children. If they are interested in farm life, that's a good sign.

Lean can take a farm in many directions. Most of this book has focused on lean as a tool to shore up a farm's bottom line and keep farms viable. The benefits of lean, however, can extend way beyond profit.

The Next Step: Your Own Farm

We're at the end of our lean journey. We've come a long way. This book started with a discussion of our farm's chaos and our overwork, explaining how we didn't have systems in place to leave the farm for even short periods of time. Then it showed how we applied a new system from a faraway place (Japan) and from a very different kind of operation (manufacturing) to parse value from waste to increase our profit margins and make our farm viable. Lean is a process, not a destination. If you visit our farm, you won't see perfection; you'll see a work in progress.

Our primary motive was simple. We wanted to stay home and make a living off our farm. Profit is part of the mix, because if we are not growing crops that produce margins, then we can't stay on the farm and keep working on our goals. Lean business sense is part of our farming.

Our inspiration to apply lean didn't come from another farm. It came from a tour of a factory. Perhaps, after fighting with so many changing variables—from the weather to finicky plants—we'd become jealous of steel and aluminum, of nuts and bolts, things that don't constantly morph and throw off plans.

But in the end, it wasn't the factory that truly inspired us. It was the ideas: the idea that a farmer could create food that had real value for real customers; the idea that work could flow smoothly, without bottlenecks and waste; the idea that farms didn't have to keep getting bigger to grow more profitable, that it might be possible to transform from turtle to rabbit rather than turtle to elephant.

Pablo Picasso is widely quoted as saying, "Good artists borrow. Great artists steal." And so we have stolen ideas used in industries all over the world to make our own little farm, with our unique set of goals, a place that sustains us. The system Taiichi Ohno used to transform Toyota can transform your farm, too. It is very doable. You'll need a few supplies—perhaps a few whiteboards, magnets, and tool hooks—and energy to reorganize and refocus. The result is higher production and less waste. When used with care, the system need not replace your core values, such as staying small or low-tech. It will, however, leave you with a more pleasant place to live and farm *well*.

3/6   "Clay Bottom"   EJG

Glossary of Japanese Terms

Gemba. The people on the shop floor or the shop floor itself. In lean, the workers play an important role in continuous improvement because they are close to waste.

Genchi genbutsu. Go and see for yourself to thoroughly understand a situation. Genchi means "go to the actual place," and genbutsu means "observe the actual product, process, or service." In lean operations, managers are encouraged to go to the source and rely on personal observation.

Heijunka. Production leveling or "leveling the load" in order to avoid rushes, which overburden people and equipment, and down time, which underutilizes fixed costs like buildings and equipment. Lean managers practice heijunka in order to create smoother and more predictable work environments, on the theory that fluctuations in production always increase waste.

Kaizen. Continuous improvement. The goal of kaizen activities is to discover improvements and banish waste until a firm achieves zero waste production. That goal might never be attained, but it still provides inspiration to improve.

Kamishibai. A form of storytelling using scrolls and illustrations performed by 12th-century Buddhist monks in Japan. Managers at Toyota use a version of the practice to instruct workers and to perform audits through visual systems. Kamishibai as applied to lean involves color-coded cards placed on a board.

Kanban. A replacement signal. For example, when milk was delivered in bottles, empty bottles set outside the door functioned as a kanban, a cue to replace. In lean operations, a kanban assists with just-in-time delivery of supplies and parts.

Kata. A repeated, ingrained action that has become second nature, like movement sequences in the martial arts.

Muda (Type 1). Waste in the form of steps that lead up to value but that do not actually create value. Examples on farms include time spent installing irrigation systems or rolling up curtains to vent greenhouses. Minimize these actions.

Muda (Type 2). Pure waste in the form of unnecessary steps that do not add value. Examples include overpackaging and overproducing. Eliminate these actions as a first step to cutting out waste.

Mura. Waste in the form of unevenness in production and sales, which creates erratic, unpredictable work environments.

Muri. The waste of overburdening people or equipment. Muri leads to equipment failure, poor work, injury, and burnout.

Poka-yoke. Mistake-proofing. Poka-yoke systems detect errors before they cause defects. The goal is to spot problems as soon as they occur, before your customers do.

Seiketsu. To standardize and make part of everyday work the practices of seiri, seiton, and seiso. The fourth step of the 5S organization system.

Seiri. To sort. In lean, seiri means getting rid of everything not absolutely necessary for current production. The first step of the 5S organization system.

Seiso. To "shine" by cleaning a space or installing lights to make it brighter. Lean workplaces practice seiso to increase quality and uncover hidden waste. The third step of the 5S organization system.

Seiton. To set in order. In lean, seiton often means storing tools at eye level, close to their point of use. The second step of the 5S organization system.

Shitsuke. To sustain the 5S organization system through self-discipline and regular audits—the final step of the system.

Resources for Further Study

RECOMMENDED READING ON LEAN MANAGEMENT

I recommend the books below for those who want to delve deeper into lean concepts. They are listed in order of importance for us as we have applied them on our farm.

Toyota Production System: Beyond Large-Scale Production, by Taiichi Ohno (Productivity Press, 1988).

Taiichi Ohno's Workplace Management, by Taiichi Ohno (McGraw-Hill, 2013).

Lean Thinking: Banish Waste and Create Wealth in Your Corporation, by James P. Womack and Daniel T. Jones (Free Press, 2003).

5S for Operators: 5 Pillars of the Visual Workplace (for Your Organization), by Hiroyuki Hirano (Productivity Press, 1996).

The Toyota Way, by Jeffrey Liker (McGraw-Hill, 2004).

The Machine That Changed the World, by James P. Womack, Daniel T. Jones, and Daniel Roos (Simon and Schuster, 1990).

The Lean Startup, by Eric Ries (Crown Business, 2011).

The Lean Six Sigma Pocket Toolbook, by Michael L. George, David Rowlands, Mark Price, and John Maxey (McGraw-Hill, 2005).

RECOMMENDED WEBSITES FOR LEAN MANAGERS

The websites below offer blog posts on lean, descriptions of workshops and seminars, suggestions for further reading, and supplies for organizing.

Lean Enterprise Institute: www.lean.org
Leadership Services: www.bill-waddell.com
Manufacturing Leadership Institute: www.idatix.com
The 5S Store: www.the5sstore.com
The University of Michigan Lean Program: www.isd.engin.umich.edu

RECOMMENDED READING ON FOOD AND FARMING

The sources below, alphabetized by title, are books that we pull off our shelves most frequently to shore up our farming skills or inspire us to keep going.

Gaining Ground: A Story of Farmers' Markets, Local Food, and Saving the Family Farm, by Forrest Pritchard (Lyons Press, 2013).

Growing for Market. A magazine for market growers.

The Hoophouse Handbook: Growing Produce and Flowers in Hoophouses and High Tunnels, edited by Lynn Byczynski (Fairplain Publications, 2014).

Market Farming Success: The Business of Growing and Selling Local Food, by Lynn Byczynski (Chelsea Green, 2013).

The Market Gardener: A Successful Grower's Handbook for Small-Scale Organic Farming, by Jean-Martin Fortier (New Society Publishers, 2014).

The New Organic Grower, by Eliot Coleman (Chelsea Green, 1995).

The Omnivore's Dilemma, by Michael Pollan (Penguin, 2006).

The Unsettling of America, by Wendell Berry (Counterpoint, 1996).

The Winter Harvest Handbook: Year-Round Vegetable Production Using Deep Organic Techniques and Unheated Greenhouses, by Eliot Coleman and Barbara Damrosch (Chelsea Green, 2009).

Notes

INTRODUCTION

1. Taiichi Ohno, *Toyota Production System: Beyond Large-Scale Production* (Portland, OR: Productivity Press, 1988), ix.
2. Ohno, *Toyota Production System*, 19.
3. Jeffrey K. Liker, *The Toyota Way: 14 Management Principles from the World's Greatest Manufacturer* (New York: McGraw-Hill, 2004), 4.
4. James P. Womack and Daniel T. Jones, *Lean Thinking: Banish Waste and Create Wealth in Your Corporation* (New York: Free Press, 2003), 10.
5. Peg Herring, "The Secret Life of Soil," Oregon State University Extension Service, February 12, 2010, http://extension.oregonstate.edu /gardening/secret-life-soil-0 (accessed December 14, 2014).
6. Bill McKibben, *Oil and Honey* (New York: Times Books, 2013), 241–242.
7. Wendell Berry, "Good Stewards," *Orion*, May/June 2014, 26.
8. Lauren Markham, "The New Farmers," *Orion*, November/December 2014, 22.
9. USDA, "Farm Size and the Organization of U.S. Crop Farming," August 2013, http://www.ers.usda.gov/media/1156726/err152.pdf (accessed December 22, 2014).
10. Lydia DePillis, "Farms Are Gigantic Now. Even the 'Family-Owned' Ones," *Washington Post*, August 11, 2013, http://www.washingtonpost .com/blogs/wonkblog/wp/2013/08/11/farms-are-gigantic-now-even -the-family-owned-ones/ (accessed December 22, 2014).

CHAPTER 1: EVERY TOOL IN ITS PLACE

1. Taiichi Ohno, *Taiichi Ohno's Workplace Management: Special 100th Birthday Edition* (New York: McGraw-Hill, 2013), 111.
2. Ohno, *Workplace Management*, 112–113.

CHAPTER 2: FARM FOR YOUR CUSTOMERS: PRECISELY IDENTIFY VALUE

1. Jeffrey K. Liker, *The Toyota Way: 14 Management Principles from the World's Greatest Manufacturer* (New York: McGraw-Hill, 2004), 230.
2. James P. Womack and Daniel T. Jones, *Lean Thinking: Banish Waste and Create Wealth in Your Corporation* (New York: Free Press, 2003), 316.
3. Womack and Jones, *Lean Thinking*, 17.
4. Womack and Jones, *Lean Thinking*, 18.
5. Jasmina Dolce, "Innovation with a Smile," *Big Grower*, March 2013, 8.
6. Grant Achatz and Nick Kokonas, *Life, on the Line: A Chef's Story of Chasing Greatness, Facing Death, and Redefining the Way We Eat* (New York: Gotham Books, 2011), 291.
7. Achatz and Kokonas, *Life*, 161.
8. Achatz and Kokonas, *Life*, 162.

CHAPTER 3: LEARN TO *SEE* VALUE

1. James P. Womack and Daniel T. Jones, *Lean Thinking: Banish Waste and Create Wealth in Your Corporation* (New York: Free Press, 2003), 94.
2. Womack and Jones, *Lean Thinking*, 316.

CHAPTER 5: FLOW I: TOOLS TO ROOT OUT FARM PRODUCTION WASTE

1. James P. Womack and Daniel T. Jones, *Lean Thinking: Banish Waste and Create Wealth in Your Corporation* (New York: Free Press, 2003), 52.
2. Womack and Jones, *Lean Thinking*, 43.
3. Eliot Coleman, *The New Organic Grower* (White River Junction, VT: Chelsea Green, 1995), 193–194.
4. Taiichi Ohno, *Taiichi Ohno's Workplace Management: Special 100th Birthday Edition* (New York: McGraw-Hill, 2013), 40.
5. Ohno, *Workplace Management*, 41.
6. Womack and Jones, *Lean Thinking*, 22.
7. Womack and Jones, *Lean Thinking*, 22.
8. Coleman, *New Organic Grower*, 166–167.
9. Taiichi Ohno, *Toyota Production System: Beyond Large-Scale Production* (Portland, OR: Productivity Press, 1988), 123.
10. Ohno, *Toyota Production System*, 121.
11. Ohno, *Toyota Production System*, 56.

12. James P. Womack, Daniel T. Jones, and Daniel Roos, *The Machine That Changed the World* (New York: Simon and Schuster, 1990), 80.

13. Gerson Cortez, "Lean Is Big When It Comes to Supply Chain," *Greenhouse Management*, May 5, 2014.

14. Cortez, "Lean Is Big."

CHAPTER 6: FLOW 2: TOOLS TO ROOT OUT FARM MANAGEMENT WASTE

1. Taiichi Ohno, *Taiichi Ohno's Workplace Management: Special 100th Birth Edition* (New York: McGraw-Hill, 2014), 44.

2. Ohno, *Workplace Management*, 44.

3. Ohno, *Workplace Management*, 26.

4. Ohno, *Workplace Management*, 137–138.

5. James P. Womack, Daniel T. Jones, and Daniel Roos, *The Machine That Changed the World* (New York: Simon and Schuster, 1990), 79.

6. Womack, Jones, and Roos, *Machine*, 79.

7. Blurb for Eliot Coleman and Barbara Damrosch, *The Winter Harvest Handbook: Year-Round Vegetable Production Using Deep-Organic Techniques and Unheated Greenhouses* (White River Junction, VT: Chelsea Green, 2009).

8. Ohno, *Workplace Management*, 47.

CHAPTER 7: LEAN FARM SALES: ESTABLISH PULL, DON'T PUSH

1. Taiichi Ohno, *Taiichi Ohno's Workplace Management: Special 100th Birthday Edition* (New York: McGraw-Hill, 2013), 67–68.

2. Ohno, *Workplace Management*, 68-69.

3. Ohno, *Workplace Management*, 68–69.

4. Ohno, *Workplace Management*, 68.

5. Ohno, *Workplace Management*, 68.

6. Taiichi Ohno, *Toyota Production System: Beyond Large-Scale Production* (Portland, OR: Productivity Press, 1988), 8.

7. Ohno, *Toyota Production System*, 9.

CHAPTER 8: CONTINUOUS IMPROVEMENT (*KAIZEN*)

1. Taiichi Ohno, *Taiichi Ohno's Workplace Management: Special 100th Birthday Edition* (New York: McGraw-Hill, 2013), 1.

2. Ohno, *Workplace Managemen*, 142.

3. Ohno, *Workplace Management*, 143.

4. James P. Womack and Daniel T. Jones, *Lean Thinking: Banish Waste and Create Wealth in Your Corporation* (New York: Free Press, 2003), 323.

CHAPTER 9: RESPECT FOR PEOPLE: LEAN AND FARM STAFF

1. Jeffrey K. Liker, *The Toyota Way: 14 Management Principles from the World's Greatest Manufacturer* (New York: McGraw-Hill, 2004), xv.

2. Liker, *Toyota Way*, 4.

3. Taiichi Ohno, *Taiichi Ohno's Workplace Management: Special 100th Birthday Edition* (New York: McGraw-Hill, 2013), 98.

4. Jasmina Dolce, "Innovation with a Smile," *Big Grower*, March 2013, 4–7.

5. Dolce, "Innovation," 4–7.

CHAPTER 10: LEAN APPLIED AT CLAY BOTTOM FARM: TEN SPECIFIC CASES

1. Taiichi Ohno, *Taiichi Ohno's Workplace Management: Special 100th Birthday Edition* (New York: McGraw-Hill, 2013), 35.

2. Ohno, *Workplace Management*, 37.

3. Henry David Thoreau, *Journal* (Boston: Houghton Mifflin Co., 1906), 160, http://walden.org/documents/file/Library/Thoreau/writings/Writings1906/15Journal09/Chapter5.pdf (accessed December 20, 2014).

CHAPTER 11: THE LEAN FARM START-UP

1. Malcolm Gladwell, *Outliers: The Story of Success* (New York: Little, Brown, and Company, 2008), 41.

2. Wendell Berry, *The Unsettling of America* (Berkeley: Counterpoint, 1996), 22.

CHAPTER 12: THE LIMITS OF LEAN IN AGRICULTURE

1. James P. Womack, Daniel T. Jones, and Daniel Roos, *The Machine That Changed the World* (New York: Simon and Schuster, 1990), 10–11.

2. Womack, Jones, and Roos, *Machine*, 11.

3. Michael Pollan, *Omnivore's Dilemma* (New York: Penguin, 2006), 59–61.

4. Womack, Jones, and Roos, *Machine*, 11.

5. Wendell Berry, *The Unsettling of America* (Berkeley: Counterpoint, 1996), 11.

6. Jeff Inglis and Tony Dutzik, "Wasting Our Waterways: Toxic Industrial Pollution and Restoring the Promise of the Clean Water Act," Environment America Research and Policy Center, 2014, http://environmentamericacenter.org/sites/environment/files/reports/US_wastingwaterways_scrn%20061814_0.pdf (accessed January 23, 2014).

7. Robert Pear, "In Final Spending Bill, Salty Food and Belching Cows Are Winners," *New York Times*, December 14, 2014, http://www.nytimes.com/2014/12/15/us/politics/in-final-spending-bill-salty-food-and-belching-cows-are-winners.html?_r=0 (accessed December 22, 2014).

8. M. D. Holmes, M. N. Pollak, W. C. Willett, and S. E. Hankinson, "Dietary Correlates of Plasma Insulin-Like Growth Factor I and Insulin-Like Growth Factor Binding Protein 3 Concentrations" *Cancer Epidemiology, Biomarkers, and Prevention*, September 2002, 852–861; J. M. Chan, M. J. Stampfer, E. Giovannucci, P. H. Gann, J. Ma, P. Wilkinson, C. H. Hennekens, and M. Pollak, "Plasma Insulin-Like Growth Factor-I and Prostate Cancer Risk: A Prospective Study," *Science*, January 1998, 563–566; H. Yu, F. Jin, X. O. Shu, B. D. Li, Q. Dai, J. R. Cheng, H. J. Berkel, and W. Zheng, "Insulin-Like Growth Factors and Breast Cancer Risk in Chinese Women," *Cancer Epidemiology, Biomarkers, and Prevention*, August 2002, 705–712.

9. European Commission, "Report on Public Health Aspects of the Use of Bovine Somatotrophin," March 15–16, 1999, 17, http://ec.europa.eu/food/fs/sc/scv/out19_en.html (accessed April 20, 2015).

10. Food and Agriculture Organization of the United Nations (FAO), "Livestock's Long Shadow: Environmental Issues and Options," 2006, xxi, ftp://ftp.fao.org/docrep/fao/010/a0701e/a0701e.pdf (accessed April 20, 2015).

11. FAO, xxi.

12. FAO, xx.

13. US Environmental Protection Agency, "Overview of Greenhouse Gases," http://epa.gov/climatechange/ghgemissions/gases/ch4.html (accessed December 24, 2014).

14. FAO, *Livestock*.

15. Pear, "In Final Spending Bill, Salty Food and Belching Cows Are Winners."

CHAPTER 13: LEAN FOR MORE THAN PROFIT

1. Wendell Berry, *The Unsettling of America* (Berkeley: Counterpoint, 1996), 7.

2. Berry, *Unsettling*, 6–8.

3. Jean C. Buzby, Hodan F. Wells, and Jeffrey Hyman, "The Estimated Amount, Value, and Calories of Postharvest Food Losses at the Retail and Consumer Levels in the United States," USDA Economic Research Service, Economic Information Bulletin Number 121, February 2014.

4. Dana Gunders, "Wasted: How America Is Losing up to 40 Percent of Its Food from Farm to Fork to Landfill," Natural Resources Defense Council, August 2012, 4.

5. Gunders, "Wasted," 4.

6. "A Trillion in the Trough," *The Economist*, February 8, 2014, http://www.economist.com/news/united-states/21595953-congress-passes-bill-gives-bipartisanship-bad-name-trillion-trough (accessed December 22, 2014).

Index

Note: page numbers followed with p refer to photographs.

About the Author

BEN HARTMAN grew up on a corn and soybean farm in Indiana and graduated college with degrees in English and philosophy. He and his wife, Rachel Hershberger, own and operate Clay Bottom Farm in Goshen, Indiana, where they make their living growing and selling specialty crops on less than an acre. Their food is sold locally to restaurants and cafeterias, at a farmers' market, and through a community-supported agriculture (CSA) program. The farm has twice won Edible Michiana's Reader's Choice Award.

Photo courtesy of Conrad Erb.

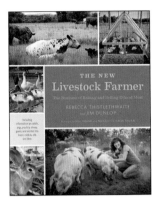
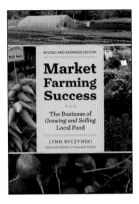
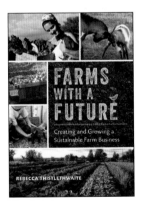
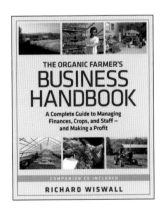

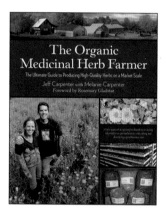
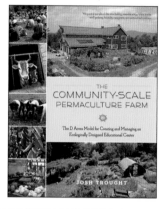

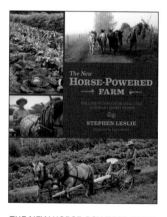